— the —

ENLIGHTENMENT
VISION

— *the* —

ENLIGHTENMENT

VISION

SCIENCE, REASON, AND THE PROMISE OF A BETTER FUTURE

STUART JORDAN

 Prometheus Books

59 John Glenn Drive
Amherst, New York 14228–2119

Published 2012 by Prometheus Books

Cover image © 2012 iStockPhoto
Cover design by Nicole Sommer-Lecht

Inquiries should be addressed to
Prometheus Books
59 John Glenn Drive
Amherst, New York 14228–2119
VOICE: 716–691–0133
FAX: 716–691–0137
WWW.PROMETHEUSBOOKS.COM
16 15 14 13 12   5 4 3 2 1

Library of Congress Cataloging-in-Publication Data

Jordan, Stuart D.
    The Enlightenment vision : science, reason, and the promise of a better future / by Stuart Jordan.
        p.m.
    Includes bibliographical references and index.
    ISBN 978-1-61614-640-5 (cloth : alk. paper)
    ISBN 978-1-61614-641-2 (ebook)
    1. Enlightenment. 2. Civilization--Forecasting. I. Title.

B802.J67 2012
001--dc23

2012031629

Printed in the United States of America on acid-free paper

To Professor Paul Kurtz, who has done more for the freethought movement in the United States and internationally than any American since John Dewey, and who has for his entire adult life fought for, argued for, and organized for the principles of the Enlightenment that are the subject of this book.

CONTENTS

7

PART 4. REASSESSING THE ENLIGHTENMENT

PREFACE AND INTRODUCTION

This book is primarily a general thesis within an even broader one. *The Enlightenment Vision* proposes a general thesis that, while the historical Western Enlightenment remains the right vision for a higher stage of human civilization, we have so far fallen short of its attainment and will not likely reach the goal soon or without further struggle. However, the broader thesis offers a prediction that this goal is likely to be attained and also exceeded in currently unpredictable ways by descendants both more intelligent and more empathetic than we are today. Arguments are given in support of these views. No rigorous proof is attempted, nor is one possible, for the future of humanity is ultimately unpredictable. What is written here can then be called an educated guess, but one that I hope to show is a reasonable one.

The Enlightenment was primarily an ethically motivated humanistic movement to improve the secular lives of people everywhere. To do so, it advocated science and reason as the means to realize that goal. I review here some of the major obstacles that have stood in the way of a more rapid acceptance of this approach to realizing a more progressive civilization, and offer some views on work in progress to overcome these inertial and reactionary tendencies.

There are three assumptions that underlie these two coupled theses. The first states that all reality external to the human mind, and arguably the human mind and thoughts as well, are part of a natural order amenable to scientific investigation. The second assumption asserts that biological evolution has produced a human species endowed with a strong genetic will to live and to constantly explore ways to expand its knowledge and opportunities for realizing and expanding its potential. The third assumption regards science as the best way to obtain reliable knowledge of the natural order, including, eventually, more reliable knowledge of ourselves.

I call these "assumptions," but based on applying the reliable evidence and rational thought held critical for progress by Enlightenment thinkers, strong arguments can be made in support of each. Thus my outlook is entirely secular. Acknowledging that there are others who would still challenge one or more of these assumptions and claim another order of reality, I welcome their outlook as long as it is a tolerant one. There are religious as well as secular humanists, but my interests and this book are restricted to the secular life that all people can experience as "real." On the basis of these three assumptions, the broader thesis claims that our descendants are virtually assured of survival and the achievement of a more satisfactory secular life, because of this built-in genetic will to live and to thrive, *and* because natural science provides the most reliable (i.e., useful or workable) knowledge we can obtain of the natural order.

Nevertheless, the survival of our descendants to enjoy more fulfilling lives offers no guarantee that this future state will be reached soon or easily. Several of the following chapters will argue that the progress envisioned is sure to meet deeply entrenched reaction and opposition. Primarily for this reason, I will argue that mindless optimism on the near term is unwarranted. I will present reasons in what follows why the broader thesis remains compelling, but only if we are willing to take the long view. When compared to the remarkable optimism of Enlightenment thinkers of the eighteenth century, there is a more sober and occasionally even pessimistic mood among many intellectuals today. I discuss this and agree with a few of their bleak contemporary prognoses, but I attribute their resulting pessimism to looking at the world over too short a time span. I will argue that taking a longer view confirms the optimism of the Enlightenment thinkers, *provided* that we *substantially* increase the time for realization of the goals.

This book is not an academic thesis. The fashion in much academic work today is to study in great detail a small, well-posed problem. There are good reasons for this, with which I am familiar in my own field of scientific specialization. It weeds out a lot of speculative nonsense. To one who might say this book is a good example, I offer the following reply. The question of successful biological descent of our distant future progeny is of great importance to most thinking people. Many of them satisfy themselves by accepting an answer on authority or providing ideas of their own. Some may even fear a not-too-distant end to the human experiment of life on earth. However, I have yet to encounter an example of any other systematic attempt to address this question and arrive at the same answer described here, for the same reasons and based on the same assumptions. Thus, while the knowledge and many of the ideas described in this book have all been developed in great detail by others, I hope the *approach* described here may prove as interesting to others as it has been to me.

Two final comments may clarify the spirit in which this book was written. I offer a number of personal views on ethical aspects of several contemporary developments, especially in the United States. I hope the reader will appreciate that any discussion of the Enlightenment necessarily entails considerations of ethics, as the foundation for the Enlightenment vision was humanistic ethics. Finally, while several examples from the history of many peoples and nations are included, a disproportionate number of the examples are American. That is simply presenting what I know best, from having spent most of my life here in the United States. The Enlightenment vision was and is global, and I hope that is apparent in what I have written.

PART 1

THE ENLIGHTENMENT

THEN AND NOW

CHAPTER 1

THE HISTORICAL ENLIGHTENMENT

THE HISTORICAL ENLIGHTENMENT VISION

The historical eighteenth-century Western Enlightenment was one of the most dramatic epochs in history. Coinciding with a rapid increase in the development of modern science and the early stages of the Industrial Revolution, the democratic ideas of this period were also critical for stimulating the American Revolution. The leading thinkers of the Enlightenment went beyond the arguably more profound theoretical thinkers of the seventeenth century to render ideas that had been developing for the three previous centuries more practicable.

The driving goal of the Enlightenment was based on humanist ethics. All the major thinkers associated with this movement aspired to improve the secular lot of humankind everywhere. Most of them were more interested in action than in abstract philosophy. In order to improve society, they wanted to change it. Some were academic philosophers as well, but their writings bearing directly on the Enlightenment were of a more practical kind, even when rooted in a strong academic tradition. A good example of such a thinker is John Locke in *Concerning Civil Government, Second Essay*.[1] Voltaire was definitely not an academic philosopher. The first sentence in his farce *Zadig* reveals his disdain for metaphysics.[2] Both authors had a powerful impact on progressive actions inspired by their writings.

The Enlightenment was the culmination of ideas that developed first in Italian city-states and eventually became widespread among educated Western Europeans during the Renaissance. Renaissance thinkers used extant classical Greek and Roman texts to emulate and eventually to modify classical ideas. In doing so, they rediscovered that secular life "in this world" could be quite attractive. This contrasted to a pessimistic view common during the Middle Ages, when many people believed that humanity had degenerated since the classical era.[3] Widespread acceptance of this dark view made it easy for these same people to believe their only hope lay in personal salvation though the Catholic Church. While historians have noted that this notion of an abrupt transition from "the Dark Ages" to a more optimistic and secular Renaissance is often exaggerated, they also have agreed that there were noticeable differences in how many people, especially the educated, felt about secular life in the later period than they had in the earlier one.[4]

The transition from the Renaissance to the Enlightenment was not a smooth one. The Italian city-states that started the Renaissance gradually succumbed to their more powerful neighbors to the north, in spite of Machiavelli's somewhat misunderstood but ruthless recipe for uniting his native country.[5] This failure plus the trial of Galileo eventually reduced the influence of the land that started the Renaissance, while others like the Spanish Netherlands carried on the tradition as the Dutch freed themselves from Spanish domination. The Protestant Reformation ensured that the sixteenth century would be a time of bitter religious strife that was highly destructive for much of Western Europe and disastrous for what would become Germany. These religious wars forced the Catholic Church that was becoming more liberal during the Renaissance (as long as its ecclesiastical authority was not openly challenged) to revert to discouraging all dissent while also implementing needed reforms. The progressive trend that led to the Enlightenment was delayed for more than a century.

Not surprisingly, a major feature of the Enlightenment was its optimism. Not only were the goals revolutionary; many of their proponents believed they were achievable in a not-too-distant future. Nicolas de Condorcet, sometimes called the noble philosopher, is famous for predicting such progress under trying personal circumstances.[6] Not every major Enlightenment figure, including Voltaire, was quite that optimistic, but the general mood of many eighteenth-century thinkers was more optimistic than that of several well-known scholars today, a subject that is discussed in detail in several following chapters. The *perfectibility of man* is a phrase often associated with the Enlightenment of this formative period. After a long period of worldly pessimism following the classical age in the West, typical Enlightenment thinkers were convinced they saw a bright light at the end of a long, dark tunnel.

Another feature of the historical Western Enlightenment was the emphasis Enlightenment thinkers placed on the use of science and reason as the best way to achieve the humanistic goal of a better world for people everywhere. While this goal was based on humanist ethics, science and reason were advocated as the means for realizing the objectives. The *Age of Reason* is the term frequently used to describe this historical period; and the scientific approach required that empirical evidence, not faith, must be combined with reason to better understand the secular world. The artist Francisco Goya, who was strongly influenced by the Enlightenment, reflected this view in one of the most famous sketches in his *Los Caprichos*.[7] Loosely translated into English, the long title reads "The sleep of reason produces terrible monsters. But imagination, combined with reason, is the mother of the arts and the source of everything wonderful." Goya was acknowledging what every artist and also what every scientist knows: that imagination is the start of the creative process. He also understood from a passionate and complex personal life that only when imagination is combined with reason, and in science relies exclusively on reliable empirical evidence, can the horrors of uncontrolled irrationality be avoided. Anyone familiar with Goya's famous Black Paintings sees this immediately.[8] The Enlightenment did not disregard the passionate and romantic sides of human life but insisted on understanding and controlling them in a rational manner.

A notable feature of the historical Enlightenment was its anticlericalism. Most Enlightenment thinkers were strongly opposed to ecclesiastical control of government and education. This is sometimes misunderstood today as opposition to all religion. While there were Enlightenment figures who were opposed to all religion—as there are thinkers today—it is wrong to say that this was true of all or probably even the majority. Deism, while denying an active role for a deity in worldly events, was not atheism and was a popular religion among many Enlightenment thinkers, including Voltaire, as well as many of the authors of the Constitution of the United States. While atheism was proclaimed during the most radical phase of the French Revolution, it was not universally popular among the educated even there. Historian J. B. Bury in *The Idea of Progress* points out that among the supporters of most Enlightenment goals were some clergy, not excluding Catholic clergy whose church lost influence during what was accomplished during this historical period.[9] It would probably be more accurate to say that what most Enlightenment thinkers agreed on was institutional separation of church and state, freedom of conscience, and liberating education from a dogmatic clergy.

Understandably, religious institutions were not always pleased with a reduced role in government, and this was not restricted to the Catholic Church—a situation that persists today and is discussed in greater detail in chapters 7 and 8. Religion has assumed the role of moral provider and arbiter for "the people" through much of history, even while many philosophers were adopting a more secular view of the origin of ethical behavior, a position that is supported today by numerous studies in anthropology.[10] The notion that the gods are believed by the general populace to exist, are useful to the politicians, but are not to be taken literally by the wise goes back further than Seneca. Democrates was arguably a skeptic, and Lucretius certainly was one.[11] It is not unreasonable to assume that every historical society had thinkers who were skeptical of the more dogmatic priestly pronouncements. Yet these skeptics were always vulnerable to religious zealots. Edward Gibbon famously argues in *The Decline and Fall of the Roman Empire* that civic virtue suffered at the hands of the rising Christianity, some of whose adherents he claimed were often crazy, believed the world would end soon, and even set fire to buildings in Rome to accelerate the process.[12] Nevertheless, the secular spirit continued to develop in the West, in spite of long historical intervals dominated by otherworldly concerns or religious wars.

Before devoting the rest of this book to an assessment of the Western Enlightenment, a brief digression may be needed to avoid confusion with another usage of the term *enlightenment*, which has an entirely different historical meaning in the countries of East and South Asia. Although European proponents of the Enlightenment intended their usage of the term to have universal applicability, other thinkers from Asia had developed their own concept of enlightenment that long predated the Western one. Since many contemporary Western students have become familiar with the Eastern usage, and many Eastern students of science in particular have adopted many features of the Western one, it is useful to note the differences.

The more recent Western Enlightenment has from the start been primarily concerned with the welfare of society through improving the lot of individuals. The Eastern usage has been more concerned with the spiritual development of individuals, though the consequences for society are not ignored, especially in the Hindu caste system and the Hinayana Buddhist tradition.[13] Another major difference is the attitude toward secular progress. While not necessarily a religion in the Western sense of the supernatural, these Eastern religious philosophies reveal a generally deeper resignation to fate. Though they may have never become serious competitors in the West to Western religious and philosophical ideas, some German philosophers in the nineteenth century found these Eastern ideas compelling.[14] Pragmatic action-oriented Western thinkers dominated during the Enlightenment and have not been generally receptive to these Eastern ideas since, with notable exceptions. They thought human secular progress was possible, and they were and remain today determined to achieve it.

This does not mean there is no interest in traditional East and South Asian ideas in Western thought. Nor is the reverse true. Just as the West is discovering the cultural depths of these once-misunderstood civilizations, East and South Asia are rushing to embrace science. This is encouraging, since the Western Enlightenment goals were broadly humanistic in principle and global in their intended reach. The synthesis of the best in East and West is something many scholars today think is a good thing, and is a major step toward a more civilized world. Perhaps Rudyard Kipling was wrong to say the twain shall never meet.[15] However, with this brief digression, this book will now concern itself entirely with the Western sense of the Enlightenment.

A more detailed treatment of these non-European philosophies and how they relate to the (Western) Enlightenment is given by humanist author Bill Cooke in *A Wealth of Insights: Humanist Thought since the Enlightenment.*[16]

THE ENLIGHTENMENT IN EUROPE

Many Western European nations produced thinkers who contributed to the Enlightenment, but most prominent for the ideas that became widespread in the eighteenth century were France and Great Britain. Scotland as well as England contributed, with Scotland enjoying a Scottish Enlightenment associated with David Hume, Adam Smith, and a few enlightened clergy. Nevertheless, France is usually the first country that comes to mind when the Enlightenment is mentioned, and eighteenth-century Paris attracted more thinkers associated with it than any other city in Europe. This is in part thanks to Voltaire, who spent several years in England, came to know many distinguished scholars there, and returned to France greatly impressed with the empirical tradition in English thought, contrasting it to the more theoretical tradition he found in France. Philosophers and several monarchs in other European nations also contributed to or attempted to implement provisions of the Enlightenment. All educated Europeans became enthralled with or threatened by the Enlightenment during the eighteenth century, depending on whether they were progressives or traditionalists.

The distinguishing feature of the Enlightenment in England and Scotland was its practical approach to problems of society, government, and science, often referred to as *empiricism,* a term used in both formal philosophy and in a less precise manner to denote an experimental approach to problem solving to see what works. The pre-Enlightenment works of Francis

Bacon and Thomas Hobbes were antecedents, but both were revolutionary for their time. Bacon argued the empirical case for science at a time when "natural theology" often still claimed precedence in establishing the laws of nature, while Hobbes offered arguments defending a government sufficiently strong to protect the rights of all from the predations of all, though his view of the government he favored was autocratic by later democratic standards.[17]

British Enlightenment thought came into its own with John Locke's *Concerning Civil Government, Second Essay*.[18] Locke argued for a democratic system and was one of the first prominent thinkers to make a strong case for a peoples' right to alter their government by any means necessary should it prove repressive of their basic rights to "life, liberty, and property." His work was popular with liberal political leaders in England, but it also had great influence on many leaders of the American Revolution, most notably Thomas Jefferson.

The contributions of Scottish friends David Hume and Adam Smith were equally influential, in philosophy and economics, respectively. Hume reminded scientists, whom he greatly admired, that there was no "metaphysical glue" connecting cause and effect, but that our theories represented only rational conclusions from many observations and experiments. He remained skeptical of causality and induction, while noting that they seemed to work in practice. This position struck a balance that combined a surgical metaphysical skepticism with practical empiricism, but it was nonetheless proscience.[19] Adam Smith's *Wealth of Nations* offered a critique of earlier mercantilism, which overregulated markets, and advocated its replacement with free-market capitalism.[20] This libertarian view has since been challenged by many progressive thinkers today, as the world has become more complex and most products that modern people use are not produced locally. The resulting progressive critique of free-market capitalism is discussed in several of the following chapters. However, few will dispute that Smith's ideas were radical for his time and led to many improvements in economic production. Liberal Scottish Presbyterian clergy also contributed to the Enlightenment. James Madison studied in their tradition in what today is Princeton University and applied some of their democratic ideas into crafting the American Bill of Rights.[21]

The Enlightenment in France can be viewed as a discovery of the possibilities for progress offered by science and technology, as well as a reaction to the religious wars of the sixteenth century and the dogmatic hostility between Catholics and Protestants that impelled them. Though writing prior to the Enlightenment, the tolerant views of Michel de Montaigne, who lived during these wars, conveyed a spirit of toleration and reason that underlay the democratic ideals of most prominent Enlightenment thinkers.[22] They also strongly influenced Jefferson and Madison in their determination to achieve a maximum separation of potentially warring religions through an institutional separation of church and state. Writing later during the eighteenth century, Montesquieu explored the proper character of a society's laws, noting the critical role of a country's geography in determining how it could best develop—a view that bears a contemporary resonance with Jared Diamond's *Guns, Germs, and Steel* but that still fell short of geographical determinism.[23] Montesquieu was not the most influential of the philosophers in France during the historical Enlightenment, perhaps due to opposition from Voltaire, but his *Spirit of the Laws* was popular in England and had great influence on the writers of the American Constitution, as noted in the following section of this chapter.[24]

The two French thinkers probably most often associated with the historical Enlightenment were Voltaire and Jean-Jacques Rousseau. Rousseau was the maverick, if judged by the criteria of the Age of Reason, which advocated science and reason as the means of achieving the goal of a better world for people everywhere, a position that the rationalist Voltaire fully supported. Rousseau remained religious, adopted a more passionate stance toward human betterment, and wrote brilliant essays that can be, and have been, interpreted differently by different political movements due to their ambiguity. However, in at least one area Rousseau has had a lasting influence. In *Émile*, his essay on the education of children, he criticized the traditional education of his day as largely irrelevant to the real interests and proper development of normal children, who need physical as well as mental challenge, are bored with constant memorization, and resent excessive discipline.[25] Because he wrote during the eighteenth century and was influential then as well as later, Rousseau is often considered an Enlightenment thinker, but hardly a typical one.

Not surprisingly, Voltaire and Rousseau often clashed. When receiving from Rousseau a copy of his essay *Discourse on the Origin of Inequality*, rationalist Voltaire applied his famous wit, beginning his reply with "I have received, sir, your new book against the human species, and I thank you for it." His following remarks were more biting.[26] We will meet Voltaire several times later, so I will not discuss him further now, except to say that he is undoubtedly the historical figure most educated people would associate first with the Enlightenment. When Voltaire's remains were returned to Paris during the French Revolution, six hundred thousand people lined the streets of Paris to view the funeral procession.[27]

There were other noted figures of the Enlightenment in France, whose views on human progress differed from Rousseau's romanticism and Voltaire's principled pragmatism. Especially noteworthy might be the group that gravitated to the Parisian home and opulent dinners of German author Baron d'Holbach. Included in this group of friendly atheists was Denis Diderot, the person most responsible for the world's first major encyclopedia. This remarkable work included not only essays on history and politics, the arts and sciences, but also a large number of illustrated articles on the latest technologies of that day, making it extremely useful as well as generally informative. These forward-looking progressives were eclipsed in their day by Voltaire and Rousseau, but as the world slowly becomes more comfortable with atheism, their influence could grow, as suggested by author Philipp Blom in his book, *A Wicked Company*.[28]

When a revolution finally came to France, it rapidly became a highly destructive one. In *Alexis de Tocqueville*, Hugh Brogan devotes the second part of his biography to the latter part of his subject's many attempts to liberalize the often-repressive French monarchy, restored in 1815.[29] Nevertheless, historical perspective suggests that the spirit of liberty, fraternity, and equality eventually became a permanent feature of France in the nineteenth and early twentieth centuries, especially during the liberal Third Republic, and it continues today. Edmund Burke, a conservative Whig member of the British Parliament, supported the American Revolution but opposed the violence and destruction of tradition in the French one in a famous study, *Reflections on the Revolution in France*.[30] Burke granted the need for changes in a society but argued that when they occur too quickly, the people cannot adapt, and they need some traditions to cling to or chaos becomes inevitable, an argument that conservatives have made frequently through the ages. French history seems to give some weight to this argument, but the counter position must be stated too. History also suggests that no ruling class ever gives up its power voluntarily.

The violent era of the French Revolution and the Napoleonic aftermath clearly contributed to a declining enthusiasm for the Enlightenment, which many conservatives and reactionaries blamed for the excesses. The brilliant Napoleon acted in the name of the Enlightenment but ruthlessly removed or replaced several monarchs in his areas of conquest, bringing the people "the benefits of the revolution" while appropriating many local art treasures for Paris. A more romantic period in the arts and a spirit of regional nationalism that was influenced by it grew in Europe, especially after Napoleon's final defeat in 1815. The progressive spirit of the Enlightenment that began in the Renaissance seemed to have run its course once again. Perhaps it would have but for a collection of English colonies looking optimistically to the future, which had conducted a revolution inspired in part by the Enlightenment vision, but which was arguably more moderate in its immediate goals and more successful in its immediate outcome. If enthusiasm for the Enlightenment suffered from the European experience at that end of the eighteenth century, it was transmitted to and survived enhanced in early America.

ENLIGHTENMENT INFLUENCES ON THE EARLY UNITED STATES

If Europe produced the Enlightenment, the first country to apply it with spectacular success was the early United States. Historian Henry Steele Commager reviews this theme in his appropriately titled book, *The Empire of Reason: How Europe Imagined and America Realized the Enlightenment.*[31] That conditions in early America were ideal for such a development— while less so in all eighteenth-century European nations—cannot be denied. The North American English colonies of that era were largely free of the burdens of history while enjoying some of the major benefits. Among the many burdens were long-established attitudes and habits resistant to change, as well as traditional, often-autocratic governments. Among the benefits were the pragmatic and democratic ideas favored by the Enlightenment, coupled with a huge, underpopulated continent rich in natural resources. Some of the better-educated immigrants brought with them Enlightenment ideas. Many immigrants brought technical skills for developing a prosperous society. There were also theocratic and plutocratic tendencies in some of the immigrant groups, which persist in America today, but if any society in modern history was favored by circumstance to implement the Enlightenment vision, it was clearly the United States. It would not be wrong to call early America the child of the Enlightenment, for without this development in Western thought the American Revolution would more likely have failed, assumed a different form, or perhaps not been attempted at all.[32]

The United States was also fortunate in having England as a "mother country." Thanks to favorable natural circumstances, the island of Great Britain offered enough physical isolation from the Continent before the air age to suppress a major invasion (save the one in 1066 CE that succeeded). This reduced the need for a large standing army and allowed the island nation to gradually develop pragmatic traditions in thought and government. These developments made the most of geography and available resources to eventually produce one of the most successful and prosperous democracies in history. Many of these traditions were transmitted to the early English colonies in North America. A majority of the early immigrants to America were from Great Britain, and they brought many indigenous British values and skills with them. There is another side to this story that is less inspiring, but it is still hard to imagine any other nation in the world as well suited as England and its British cousins to provide the settler stock of the original colonies that became the United States.

Even during the American Revolution, many liberal Whig politicians and at least one distinguished Whig conservative, Edmund Burke, supported independence for the American colonies. Historians Henry Steele

Commager and Richard Morris review the revolutionary period in detail in *The Spirit of 'Seventy-Six*.[33] They provide excerpts from both British and American sources that reveal many remarkably balanced proceedings in both the American Continental Congress and the British Parliament during the Revolution as well as before and after it. These proceedings occurred during bread riots in London and extreme financial austerity in America. Democracy survived in both societies, a historical note probably worth recalling today.

Two different perspectives on the American Revolution illustrate the formidable body of European (including British) Enlightenment thought that inspired it and the clearly written, down-to-earth approach of its most famous pamphleteer. Contemporary historian Bernard Bailyn discusses these Enlightenment ideas in *The Ideological Origins of the American Revolution*, arguing that the European Enlightenment provided diverse reasons for breaking with more traditional forms of government, while discussing how to create a powerful new one that would not undermine the liberties of its citizens.[34] His study reveals how this process evolved over more than two decades, noting that the Enlightenment ideas that inspired it were by no means a uniform ideology. Bailyn argues this was favorable for providing the founders of the young United States with alternative views that were nonetheless all focused on the desire of their often-contentious European authors to advance human welfare.

Thomas Paine's easy-to-read *Common Sense* inspired a large number of colonial readers to support the break with England, while the opening statements in his *American Crisis* may have done as much to motivate and preserve Washington's small army at Valley Forge in the winter of 1776 as any other single event of that time.[35] Talented pamphleteers who reached out to a large audience played important roles in furthering the democratic, humanistic spirit of the Enlightenment. European scholars who influenced the American Revolution and these skillful pamphleteers shared the desire to see Enlightenment ideas applied to the problems of liberty and good government. They were in that sense all practical-minded people, a quality often attributed to American culture when faced with the challenges of settling a new country. Whether and how that quality has been preserved when subjected to the responsibilities and stresses of becoming a world power is considered in later chapters.

We have already noted the influence of Locke and Montesquieu on American thinkers of our own revolutionary era. Jefferson borrowed a great deal from Locke's arguments for the right of a suppressed people to revolt, and by replacing Locke's advocacy of a people's right to life, liberty, and property, with "life, liberty, and the pursuit of happiness," he crafted the most famous sentence in the Declaration of Independence. Writers of the Constitution familiar with Montesquieu's ideas on separation of the three main branches of government incorporated his views on separation of the legislative, executive, and judicial branches of government into that document. This structure remains today in American state governments as well as the federal one.

Alexis de Tocqueville may not be considered by all scholars as a typical Enlightenment thinker, since his two most famous works were written well after the Romantic Reaction had eclipsed the Age of Reason associated with the historical Enlightenment. Nevertheless, in his obvious, if astutely critical, enthusiasm for democratic government and personal liberty in *Democracy in America*, Tocqueville was greatly influenced by many Enlightenment ideas and was no friend of tyranny.[36] While a devout Catholic, his comments on how the Catholic religion might thrive in America are instructive. In a carefully worded statement, Tocqueville recommended that the Catholic Church would do well in the United States . . . if the clergy stayed out of politics and the private lives of the people. We return to this interesting view in chapters 7 and 8.

Tocqueville's writing also reflects the cautious side of American revolutionary enthusiasts who struggled with the natural conflict that inevitably arises between liberty and equality. He expressed a concern that if the people should succumb en masse to a powerful movement strong enough to override established personal liberty, American democracy could degenerate into a tyranny of the majority, a potentially pernicious consequence of pure egalitarianism. Recognizing this possibility led to skepticism among some authors of the Constitution over how much the people should be trusted to govern themselves.

We still do not know in detail what occurred behind the closed doors of the Constitutional Convention in Philadelphia. We do know that through rejection of pure egalitarianism and a final agreement to provide a subsequent Bill of Rights to protect personal liberties, a representative form of government was created that struck a balance between extremes on several issues of power and human rights, and it has functioned successfully with episodic spasms for well over two centuries. In a recent biography of Tocqueville, Hugh Brogan, after assessing Tocqueville's shortcomings, offers a view shared by many progressive as well as numerous conservative scholars today. Brogan provides a final assessment of the book by this complex, courageous French aristocrat. "*Democracy in America* is the greatest book ever written on the United States."[37] That is high praise for a work that incorporates both admiration for America and a sober warning.

THE HUMANISM OF THE HISTORICAL ENLIGHTENMENT

The humanist tradition in the West has a long history, even if the word *humanism* as understood today was invented in the nineteenth century, which is asserted by author John Hale in *The Civilization of Europe in the Renaissance*. Hale claims that Renaissance writers used *humanism* to describe "the conditioning of ideas that drew on a knowledge of classical antiquity."[38] He adds that the term *humanism* was actually inspired by a fifteenth-century Italian scholar who devised a curriculum of classical studies, *litterae humanioris*, a term that is still used for that purpose at Oxford University today. Enlightenment thought has given us the modern sense of humanism, which is now regarded as an outlook that emphasizes the welfare of human beings in the secular world. This does not deny the existence of religious humanism to those religious people who remain tolerant and who support the idea of human progress to a better secular life. However, many self-identified humanists today regard themselves as secular humanists, and some, like distinguished science educator Richard Dawkins, proudly proclaim atheism.

These historical developments inspired American humanist leader Paul Kurtz to apply the term *secular humanism* to his pragmatic philosophy based on Enlightenment principles. Not surprisingly, religious reactionaries in America and elsewhere attacked "the religion of secular humanism," and Kurtz wrote *In Defense of Secular Humanism* in response.[39] An excellent review of how humanist ideas developed from the Renaissance up to 1984 is found in Alan Bullock's *The Humanist Tradition in the West*.[40] Bullock notes that the high watermark for humanist ideas may have occurred during the Enlightenment, and he offers some sobering observations in his last chapter, titled "Has Humanism a Future?" He maintains a hopeful note but with reservations based on several modern developments that will be examined further in this book.

I begin my own examination of Bullock's query on the future of Enlightenment humanism in the next chapter. Two centuries of historical developments since the Enlightenment, and especially since the two World Wars of the twentieth century and the Cold War that followed, have significantly reduced the optimism of many thinking people for rapid human betterment. This has occurred notwithstanding spectacular progress in all the natural sciences, in physical technology, and in medicine. In some cases, even the idea of progress, if by that is meant *societal progress*, has been called into question. Traditional religion, defined as faith in the supernatural, which many Enlightenment thinkers thought would rapidly disappear with the rise of science, exhibits renewed vitality in many parts of the world today, including the United States. Nor can a resumption of interest in traditional religion be declared moribund even where it is currently weak, if favorable conditions for secular life cannot be maintained. Much of the rest of this book will assess the consequences of developments since the eighteenth century, providing the basis for reassessing the basic principles and goals of the historical Western Enlightenment.

CHAPTER 2

STATUS OF ENLIGHTENMENT GOALS TODAY

VARIETIES OF SOCIETAL PROGRESS

We have noted the optimism of the European philosophers and the leaders of the American Revolution regarding the rapid improvement in the lot of humankind, guided by humanistic ethics and implemented by scientific discoveries and the application of reason. This chapter asks how well that optimism has been justified by events since the end of the eighteenth century. Recalling that the overriding goal was the development of a better secular life for people everywhere, we ask what progress has been made toward achieving that goal, how permanent that progress has been, and in what ways the goal has proved just out of reach or has not been achieved at all.

There has undeniably been great progress in both science and science-based physical technologies and medicine. Where adequate resources and relatively stable societies have developed, the economic standard of living and the average longevity have increased substantially over the last two centuries. This situation is particularly apparent in the United States, most of Europe, Japan, many parts of East and South Asia, and in local regions over the rest of the world. While further progress is needed in many areas, these consequences of modern science, technology, and medicine are spectacular human achievements, especially considering that they have occurred over no more than three or four contemporary human life spans in the societies mentioned above.

There has also been a substantial growth in the number of democratically elected governments over the last two centuries, especially since the defeat of the major authoritarian powers in the twentieth century. Since democracy is by definition rule by the people, if usually indirect, this constitutes a major advance in human rights. Some of these democracies have proved short-lived for reasons we review in chapter 8, but even many of those failed democracies have slowly moved back toward governments more responsive to the people. As these words are written (in 2011), democratic movements have taken to the streets in Southwest Asian and North African Islamic nations. While the outcome is currently difficult to predict—and elements hostile to political democracy are present in all these countries, as elsewhere—there is no doubt that large numbers of people in the Islamic world are no more attracted to a theocratic oligarchy than they are to a repressive fascist-style regime.

There are, however, other measures of human welfare that force us to pause before declaring the Enlightenment vision a clear success, largely achieved. In addition to scientific and technical progress and the societal benefits that have flowed from it, there are other measures of societal progress that have an obvious impact on personal happiness and a sense of general well-being. Poverty and significant human-rights abuses are still present in most societies and are common in some. Racism and sexism remain widespread. Militant nationalism and religious fanaticism continue to erupt, with the danger of destabilizing entire regions; and prosperous democratic nations like the United States are not immune to their effects. Terrorism continues to plague parts of the world, and modern Internet technologies have so far made terrorism easier to implement everywhere. As the threat of world war decreases, the destructive potential of weapons of mass destruction, not all of them nuclear, seems to increase in inverse proportion; while small-scale wars continue, typically producing more civilian casualties than military ones. Environmental devastation proceeds apace and continues to be difficult to regulate, especially when industries become global in scope, command large purses for "public education," and strive for increasing control of the public media. No sensible person today thinks there exists an effective international regulatory system to address many of these problems, yet all educated people recognize that most of them are planetary in extent.

These problems and their consequences for people's lives require a definition of societal progress that incorporates them in an assessment, along with the great benefits that modern societies have derived from the rapid developments of science, physical technologies, and medicine. If we return to the basic, humanistic, ethics-driven goal of the Enlightenment, to develop a world that offers human beings everywhere more fulfilling secular lives through a proper application of science and reason, it becomes clear that societal progress has failed to realize that goal in many areas. *I will define* societal progress *as how well the broad goals advocated by Enlightenment thinkers and stated here have been reached by us today.* In some cases, we can grant great progress. In others, we have clearly failed to date. This raises the questions of whether the original goals were too optimistic for human beings, or whether eighteenth-century thinkers were unrealistic in their assessment of the timescale for their realization.

Before assessing societal progress, the problems described above must be confronted, individually and together. These are the issues we consider briefly in this chapter, and they will be discussed in more detail in several chapters that follow. There, the case will be made that, in agreement with many scholars, the human race has yet to grasp the full consequences of the rise of modern science and the Industrial Revolution and has certainly not learned how to manage these consequences well. The modern "world system" is seen to be running somewhat wild, often with no one in effective control. This situation is clearly dangerous, since widespread injustice and anger, violence, unregulated greed, and environmental degradation continue. Imagining a desirable future state is an essential first step to a better world for people, but assessing the starting point is the next essential step, or the visionary effort can prove pragmatically useless. A world-class sprinter cannot win a race if he fails to locate the starting line.

SOCIETAL PROGRESS TO DATE: SUCCESSES AND FAILURES

Eighteenth-century thinkers were sufficiently realistic to know the knowledge base for realizing their ambitious goals was not available at that time. However, most of them had enormous faith in the power of human reason to overcome the obstacles to progress. Combined with increasing scientific knowledge of the world and of medical procedures for curing the diseases that caused most early deaths in their era, they anticipated a relatively rapid increase in the move toward a better secular human world. A case for cultural bias can be made. The traditional Enlightenment was a product of Western civilization and probably contained typically modern "Western" biases, including a preference for individualism and a sometimes unbridled optimism, if not always in the secular realm then in the spiritual one. Outside of a remote Nirvana, no enduring Asian philosophy has ever advocated the perfectibility of man, and none to my knowledge has done so for our secular existence. European knowledge of the great Asian civilizations was still limited at the time of the Enlightenment. On the cautious side, there were famous Western thinkers like Hume, who recognized that reason could be the slave of the passions, and Voltaire, who declared it the weakest of the passions, possibly anticipating modern discoveries in neurophysiology that reason and emotion cannot be entirely separated. Nevertheless, until the clearly emotion-driven romantic reaction to the so-called Age of Reason superseded Enlightenment thought, many visionary Western thinkers of the eighteenth century viewed reason as akin to a goddess who, like a rational Athena, would lead humankind invariably forward to a better age, and would do so rather quickly.

To better determine how well the Enlightenment vision has been realized today, we need to go beyond the broad vision and ask what particular achievements were included in that vision; then ask if those achievements have been realized. This will address some of the same broad issues raised in the previous section but will provide a more detailed basis for their assessment. Where clear failures have occurred to date, this will also aid us in identifying major obstacles to further progress.

We can start with the economic well-being of people around the world. This breaks down into several important sub-issues, many too detailed to be considered here. Nevertheless, statistics on gross domestic product collected over the past half century reveal both a general rise in productivity per capita over much of the globe, even when adjusted for inflation, with some fluctuation during periods of significant recession. This is undoubtedly one of the greatest triumphs of modern science and technology, especially as applied to agriculture and the worldwide distribution of food. Thus it seems that more people are enjoying a higher physical standard of living today than ever before, both in absolute numbers and even as a percentage of a still-growing global population.

Unfortunately, when the distribution of this global wealth is considered, the picture is less encouraging. The control of the means of production remains in the hands of a tiny fraction of the world's population. While the competence of most of these individuals cannot be doubted, their control by regulatory mechanisms is often stoutly resisted and is relatively limited, especially when the same individuals control much of the public media that disseminates information on their activities. This well-known structural problem posits that unregulated or poorly regulated mature (large corporate) capitalism is subjected to processes that inexorably drive such an economic system to monopoly, imperialism, and war, often in that order.[1] History affords many examples, and the argument is compelling, even if an equally compelling argument can be made that "demonizing the rich" is the wrong approach to a solution. The leader of a large corporate enterprise must be exceptionally visionary and skillful if he is to both apply humanistic ethics and also thrive under the competitive conditions of an ineffectively regulated mature capitalistic economy. In effect, the ethical, conscientious, well-educated corporate president/CEO can easily become a prisoner of this system or else lose his job.

Thus we conclude that the solution to creating a more just distribution of wealth is a harder problem than merely replacing one set of business leaders with another who might promise major structural changes but then fail to deliver them. I have little doubt that the top executives at ExxonMobil are highly qualified leaders of modern business practice and applied technology. I have equally little doubt that some of them have approved measures to confuse the public about the dangers of greenhouse gases and their demonstrable role in producing climate change.[2] We can conclude that the system itself, when free to operate with little public constraint, is less ethical than the people trapped within it . . . as effectively as carbon dioxide in the atmosphere traps heat through the planetary greenhouse process. Correcting this structural problem in a modern capitalist economic system will be a herculean task.

Thus when socioeconomic justice is added to the picture of technological progress, the real progress made in the higher standard of living more people enjoy today is somewhat counterbalanced by major inequities in the distribution of wealth. This problem is becoming particularly acute in the United States today. Numerous studies demonstrate convincingly that as government regulation of large financial institutions and corporations has been reduced since 1980, the ratio of income of the wealthiest people in the country to that of those in all less-wealthy groups grew very rapidly to one of its highest levels in the nation's history. Talent should be and is richly rewarded in America, but it is not inappropriate to ask if the rewards, especially of inherited wealth, have become excessive.

Advancing human rights was another goal of the traditional Enlightenment vision. Whether *liberté, égalité, fraternité,* or life, liberty, and the pursuit of happiness, human rights was a major objective of Enlightenment activism. It remains so today among progressive thinkers and also among many sober-minded conservatives. But how well have human rights actually fared over the past two centuries? As with economic progress and socioeconomic justice, the picture is mixed.

Two great achievements are undeniable: the growth of democratic movements and governments and the final end of human slavery as a government-sanctioned institution. We can also note solid progress in many other areas, as racism and sexism have significantly receded in many parts of the contemporary world. Nevertheless, this progress remains regional and episodic, with many examples of regression. The most optimistic statement we can make is that the worldwide trend seems to be in a progressive direction but is typically slow on the scale of a human lifetime. These questions of realizing human rights are treated in greater detail in chapter 6, particularly on the reduction of racism and the gradual, if uneven, increase in women's rights.

Militant nationalism, sometimes enhanced or even organized by religious fanatics, has been a problem since the rise of nation-states, and it remains so today. Militant nationalism runs completely counter to the Enlightenment goal of universal peaceful coexistence, a necessary condition for optimum human progress. Democracy is always threatened by militant nationalism, for all nations in the grip of military enthusiasm invariably, and often necessarily, become more authoritarian for the duration of the real or perceived crisis. American history offers no exception to this general truth, as those familiar with American Civil War history or certain actions taken by the federal government during the two world wars of the twentieth century can attest.

Militarism also provides opportunities for immense corporate profits, leading to a military-industrial complex that develops a life of its own and is difficult to ramp down when the major crisis has passed. The United States addressed this problem quickly—some would say too quickly—at the end of World War II, but it seems to have had difficulty addressing it since the collapse of the former USSR. Perhaps the most dangerous tendency of all powerful nations is promoting the pernicious myth of national exceptionalism, as opposed to the valid idea of national uniqueness. The myth of national exceptionalism encourages national populations to feel intrinsically superior to people in other countries, and it can also make them more warlike, even in peacetime. Universal Enlightenment goals can then be replaced with a narrow, nationalistic chauvinism. These issues are treated in greater detail in chapter 8.

Repressive militarism often creates the conditions that terrorism and guerilla warfare thrive upon. Military repression of a people is a known stimulant for the development of terrorist organizations and guerilla fighters, who can live off the land in supportive areas, especially in the countryside where the terrain affords additional protection. Today, in the Internet age when tight control of information is difficult to establish and people who feel oppressed may offer safe haven for terrorists and guerilla fighters, all areas of the world that harbor disaffected populations become candidates for training and equipping these nonmilitary combatants. Terrorism and guerilla raids may be the angry response of the weak to the strong, who are often not guilty of many terrorist accusations, but it is hard to imagine extensive terrorism and guerilla warfare in a world where true socioeconomic justice would prevail. Obviously, a broad perspective is needed to assess the sum of the underlying causes of contemporary terrorism. Suppression is often required, but suppression alone is clearly not enough, and the scourge continues today.

I conclude that as long as large-scale socioeconomic injustices continue among nations and no effective international authority exists to correct them, the danger of local violence and wars on many levels will continue. While the reluctance of both the United States and the former Soviet Union to engage in a mutually catastrophic exchange of nuclear weapons gives us reason to think that fear of Armageddon may induce highly antagonistic societies to stop short of global war, accidents can still happen. The message of physicists who have tried to educate the public on this issue through the *Bulletin of the Atomic Scientists* is clear.[3] If we wait long enough in a nuclear-armed world, some kind of nuclear weapon will eventually be used again, with a serious danger of retaliation in kind. While universal horror at such an attack could prevent an uncontrollable exchange, this outcome is not certain, especially when dealing with fanatics.

More likely on the shorter term is a series of local wars that small and often politically backward nations engage in for a variety of reasons, often under an autocratic leader seeking to enhance his own power or divert his people's attention from their domestic suffering. At this writing in 2011, North Korea may afford the best example. The greatest danger such small nations pose may be to draw other larger, more powerful nations into such a conflict. If a war should develop between the two Koreas, it will require skillful diplomacy on the part of both the United States and China to avoid being drawn into the conflict on opposite sides. Even the dangers of a new cold war developing between America and China cannot be ignored. There are those in the American corporate world, and probably certain militarists in China, who under certain conditions deemed favorable to them might wish to provoke one to advance their own agendas. It is hard to imagine anything more unethical, more tragic (or, speaking frankly, more stupid) than an unnecessary cold war between the United States and China, with a colossal diversion of resources that could have been used for human betterment. Only a hot war would be worse.

An encouraging study by American scholar Steven Pinker suggests that the danger of death from violent causes has dropped impressively over time.[4] This may continue if no major future war occurs in which nuclear weapons are used. However, a nuclear detonation will instantly kill orders of magnitudes more people than any other known contemporary weapon. If a significant number of nuclear weapons were used in a future conflict, Pinker's optimistic projections could still fail. Fortunately, there are still people in politics who devote great effort to addressing this danger. No person in America today has done more than fiscally conservative Republican senator Richard Lugar in trying to reduce nuclear proliferation. Yet Lugar's own political party recently seemed reluctant to support President Barack Obama (eventually doing so with some reservation) when he obtained Senate confirmation of the latest START treaty with Russia to further reduce nuclear warheads in both countries. Should a question of this magnitude become a political football?

While the risk of nuclear war is the most dramatic evidence of how far our civilization falls short of achieving the Enlightenment goal of a peaceful, planetary society, the greatest threat to global human well-being in the near future may come from a nonmilitary source, climate change. The approach to this potential worldwide disaster is insidious and gradual, though evidence for it has been accumulating over much of the previous century. Abundant observations show that climate change is already in progress. Other forms of environmental degradation have also been well documented, and many are closely correlated with the growth of the human population, and even more with certain forms of human industrial activity. There is a sad irony here. If environmental and ecological conditions would permit further population growth, most of us would welcome it. Regrettably, the same industrial activity that has improved our quality of life in diverse ways has also proved to be a primary cause of environmental degradation, along with further, possible unsustainable population growth. Even if major corrective action is taken now, harmful climate changes will continue for decades. These are now known with increasing scientific confidence to be caused largely by human production of greenhouse gases.[5]

It is hard to exaggerate the difficulty of solving this problem, for doing so involves far more than the science that is now relatively well understood. A serious effort at solution must consider the production of energy on which our economic well-being and the employment of millions of workers depend. The political aspects of such a problem invariably become turbulent. Demonizing any one group has proved to be, and will likely continue to be, unproductive. Yet, since the underlying science has become the subject of gross misrepresentation, and possibly outright lying, on the part of ideologically biased "experts," we have a situation where the most basic factor needed for solution—a scientific understanding of the problem—has become confused in the public mind. Because of the critical role science plays in providing the reliable knowledge for making major decisions a democracy depends on today, democracy itself could break down if misrepresentation of science for economic gain or political advantage continues. Readers interested in how some have attempted to deceive the public on the seriousness of the global climate-change problem in particular, and on several other science-based problems before, are encouraged to read several recent books that summarize the techniques that have used been used to confuse the public, often with success.[6] Kendrick Frazier, editor of the magazine the *Skeptical Inquirer*, has collected essays from a number of scientists and skeptics in one book, *Science under Siege.*[7] Many of these essays demonstrate that science itself is often attacked in the name of what can only be called pseudoscience.

Lurking in the background of all the unsolved problems discussed in this section is one more global problem that the human race has clearly failed to solve to date. There is no worldwide political authority adequate to regulate the major activities of powerful sovereign nations. This now also applies to the activities of many international corporations, which even major nations find difficult to regulate. The United Nations has provided excellent medical and other services to needy parts of the world, and some thinkers have expressed hope that a strengthened UN might provide the nucleus for a future effective world government.[8] However, when major issues of war and peace or economic development arise, the UN is often ignored, especially by the major powers. When the United States invaded Iraq in 2003 without UN sanction and the going got rough, the UN simply left that country. It had no hope of influencing events and found it prudent to protect its own people, some of whom were killed during that conflict. The universal goals of the Enlightenment can hardly be realized in a world where every sovereign state is a potential future rogue nation.

Effective world government may be a prospect the current stage of human civilization is simply not ready to embrace. There are many sound, pragmatic arguments for not pursuing world government too quickly. However, it is hard to think that optimum solutions to common great world problems can be achieved without it. Some of these issues will be discussed further in chapters 8, 9, and 12.

A summary of the successes and failures to advance toward a more humane world can be given, based on the previous discussion. This can be useful for identifying some common features of both the successes and the failures. The schematic summary given below is rough but revealing.

SUCCESSES	FAILURES
Increased general well-being	An unjust economic distribution of wealth
Humanistically valuable technologies	Massively destructive technologies A damaged and degraded ecosystem

MIXED

Impressive advances in democracy and human rights, with a long way still to go

Of course this summary is a huge generalization and can be challenged in many ways. For example, if America had not developed the atomic bomb, the democracies might have lost the war if Germany or Japan had succeeded in developing one. This is true, but there are always other consequences of major human decisions that can drive the application of our science and technology in dangerous directions. Thus long-term as well as compelling short-term considerations need to be applied to these decisions if we are to solve our major global problems. The decision to use the atomic bomb to end World War II may be an example of such a decision, and those who have studied the reactions of different thinkers on the question have noted a less than objective assessment on both sides of the question. Another example is the structural problem of inadequately regulated mature capitalism referred to in the previous section. It was noted there that monopoly, imperialism, and war have been a frequent consequence. A further example is war itself. Most of the world's people probably prefer peace, but if any major nation chooses the path of aggression and conquest, the rest of the world that is affected has only two choices: capitulation or war. If the aggressor is powerful, initially successful, and technically sophisticated, the incentive to develop increasingly powerful weapons is compelling. This has led to a world heavily armed with the weapons of mass destruction that we have today.

In light of these considerations, the rough summary of successes and failures given here suggests that progress toward a world in which a growing majority of the human population is enjoying a more fulfilling life is mixed. Progress toward some of the main goals proposed by Enlightenment thinkers has been achieved. Yet the problems that remain are extraordinarily complex. Scholars familiar with these problems generally agree on one thing. Simplistic, ideologically based solutions will invariably fail.

A quantitative assessment of these issues is well beyond this study, and even the general assessments made are subject to criticism from different perspectives. Measures of productivity and the distribution of goods are generally available today, and statistics on the distribution of wealth within and between most nations can be found. However, the question of people's subjective sense of the quality of their lives is much harder to assess. Common sense, supported by many studies, says that material well-being and health are major contributors to happiness, but experience demonstrates convincingly that "bread" alone is not enough for it. We are complex, restless, curious beings who become easily bored when simply well fed, clothed, and sheltered. A good case can be made that we have evolved to be excellent but restive problem solvers, because there are and always will be more problems to be solved. Furthermore, some of our activities that are beneficial under one set of conditions can become harmful under other, changed circumstances. We have already noted that some of the same technologies that provide energy to power our current American lifestyle can have harmful worldwide environmental consequences. A quantitative assessment of people's subjective sense of their well-being that would command nearly universal consent is probably not currently possible.

However, we can still try to identify some of the largest *obstacles* that make further progress toward solving major global problems difficult. These are the obstacles that, if not removed, make rapid improvement of the universal human condition exceptionally difficult. Some obstacles, if ignored, could leave our near-term descendants facing greater large-scale collective dangers than humanity has yet experienced to date.

Most people would agree that reliable knowledge is essential for solving difficult problems. That scientific knowledge is the most reliable kind (demonstrated in chapter 4) is a statement the majority of people would accept today if we restrict ourselves to the natural world. While the limitations as well as the extent of current scientific knowledge must be considered when reliable information is sought, a serious problem is encountered when traditional beliefs conflict with what science has already established to be true. This leads us to the problem of ignorance, which is the first topic treated in the next chapter. Thus, one major obstacle to further progress is widespread ignorance, along with a particularly common form of ignorance that might be called negative knowledge, namely, superstition. Ignorance is obviously a *major* obstacle to progress, and as in science, now that many of the "easier" problems have been solved, the remaining problems that we are aware of today are becoming increasingly difficult to solve. Providing effective education to the world's population to overcome widespread ignorance is such a problem.

Looking at the three areas identified above in which Enlightenment goals for human betterment have failed to produce favorable results, a strong case can be made that another major obstacle to progress affecting all three is increasingly dysfunctional institutions. A dysfunctional institution can be defined as one created to solve a legitimate problem facing a society during one period but that has blossomed into a gigantic self-perpetuating system that has developed a life of its own and a momentum to keep moving in the same direction regardless of the problems it is causing. Two examples have already been discussed. One is a colossal industrial enterprise typical of mature capitalism that has become "too big to fail" because of the immediate impact that would have on the welfare of a society's population. The other is the American military-industrial complex that helped the Allies win World War II and blocked the expansion of an authoritarian form of socialism until it finally collapsed—all measures that most of us applauded. Unfortunately, this huge enterprise now has a life of its own and seems unwilling to accept that a time might come when it should play a lesser role in this democratic society, a problem that is discussed in greater detail in chapter 8.

The following chapter will consider how ignorance, superstition, and uncontrolled technologies have become major obstacles to progress toward a more humanistic world.

CHAPTER 3

IGNORANCE, SUPERSTITION, AND JUGGERNAUT TECHNOLOGY

WIDESPREAD IGNORANCE

At or near the top of the list of obstacles to progress toward a more humanistic world is widespread ignorance. This becomes increasingly apparent the farther we look into the future. The reluctance of many people to explore the unknown, whether mentally or physically, is well-known and not always bad. Conservative inclinations provide societal stability and prevent every wild idea from inciting a gullible public. Nevertheless, humanity would probably vanish from the natural order without our mental and physical explorers. The philosophers of the Enlightenment, like the early navigators of the global oceans, were such explorers, with their eyes focused on a distant horizon and beyond. The space program of NASA is similarly motivated and has proved very popular with the American public, like its counterparts in other countries. Future progress depends critically on these ventures.

Today, explorations of any kind increasingly depend on science. Unfortunately, many of the attitudes that accompany ignorance can promote denial of well-established scientific results needed for further progress. This is particularly true in several fields of medicine, which is ironic, for the same people who insist on demonstrably outdated views of human origins that appear in ancient religious texts usually want the benefits of modern medical science as much as the rest of us. Yet the appeal of discredited religious concepts can persist in an intelligent person raised with them. The believer may be torn between honoring his traditional faith and abandoning it altogether to embrace modern science. A more common approach is to rationalize that the two are somehow not logically incompatible, or to simply not think about the incompatibility and hope to enjoy the best of both worlds. Overcoming ignorance on these issues continues to be a formidable challenge, but denial of scientific knowledge has consequences. When valuable medical research is blocked on embryonic stem cells, for example, cures for certain diseases may be postponed and lives unnecessarily lost.[1]

It might seem obvious that, once informed, everyone would understand the importance of education to overcome ignorance and support it. However, using the current situation in America as an example, this is often not true. The current American population includes many tradition-minded people who find the complex modern world confusing and/or threatening. A fraction of them are tempted to reject modernism entirely, to restore an imagined superior past, or even to await the "end times," a concept astonishing to any person with a semblance of a science education. Fortunately, the fraction of those who actually believe such bizarre notions is probably quite small, and the fraction actually preparing for it is much smaller. Nevertheless, the consequences for proper public education in parts of the country are serious and not always encouraging.

Many of the people just described hold dogmatic religious views and look with suspicion on learning beyond the basics. Some favor only a limited education for their children, wanting teachers to impart the basic mental skills necessary for survival and economic functioning, but often regarding education in science as unnecessary and education in critical thinking as "corrupting." Thus science and/or critical thinking are often taught poorly, if at all. The embarrassingly low average scores American students make on standardized tests of progress in verbal and math skills and scientific knowledge combine test results from some of the most advanced students in the world with test results from others who can barely make the grade at all.[2] The same kind of widening gap between income that has developed recently in America seems to be growing in educational achievement.

In addition, those who would restrict public schooling—or home-schooling—to reading, writing, and basic arithmetic are not the only ones who may be blocking better public education. There is another group of people whose influence on the greater society is arguably larger, and their influence on restricting public education to the most basic skills may be more subtle and even more insidious.

This second group consists of *some* individuals primarily interested in competent but compliant workers, and others who believe that only a favored few will ever be able to perform difficult tasks, even if equal opportunity and a proper education were provided for all. This attitude is not shared by many leaders of modern industry, nor does it appeal to most other Americans today who have become aware of how much early personal circumstances can influence success later in life. Notwithstanding, these conservative—or reactionary—views of humankind still persist and are often associated with concerns over the possible consequences of providing genuine education to "the people." The corresponding argument states that having too many educated people in society makes it more difficult to achieve a compliant workforce or a disciplined military. The traditional notion that the public order demands education for an elite and training for the multitudes has a long and venerable history. If the primary goals for a society are order and stability above all else, a strong case for it can still be made. Educated people are demonstrably more likely to resist indoctrination and will discern societal injustices more readily.

Thus it might seem that the enlightened self-interest of the traditional conservative would compel him to argue for restricting quality education to a minority slated for positions of leadership or high professional achievement, while providing good training in basic mental skills for all others. Not a few Enlightenment thinkers of the eighteenth century were aware of this argument, and not all of them could find satisfactory answers favoring equality of opportunity in their time. No less an advocate of liberty than Thomas Jefferson is famous for advocating government by "an aristocracy of merit." This position was a great advance over favoring hereditary aristocracies, by opening up societal leadership to people of demonstrated talent who lacked the advantages of aristocratic birth. It favored many American leaders who rose to success from the general population without the traditional advantage of birth. Barack Obama is the most noteworthy present example. However, in the absence of effective public education and the current power of money to influence how society is managed, it can still be argued that the pool of future leaders and a public capable of understanding the modern world is far more limited than it could be if universal, quality public education were available to all. Whatever the reasons, which can include a concern for the up-front cost, some American conservatives do not support it.

Given these obstacles, there is still reason to think that a better educational system for the general public may develop in the United States. The dependence of modern industrial technology on employees capable of innovative work requiring the analytical skills of an educated mind has stimulated support for quality education among the more enlightened leaders of modern industry. Enlightened self-interest has favored this support for quality education of corporate employees among their employers, some of whom might otherwise have opposed it. Science education in particular has now become popular with a large fraction of the leaders in American business and industry. Even the American military, necessarily the most hierarchical and authoritarian of all the nation's major institutions, has recognized the rapidly increasing need for smart soldiers who can think on their feet and operate weapons derived from modern, science-driven technologies. A few reactionaries of the extreme political right may still see genuine education as a devil's bargain, for the reactionary mind seldom takes well to critical thinking on the part of those who disagree with them. However, when a business leader evaluates what will improve "the bottom line," it will be difficult for him to reject quality public education where profits are likely to depend on it in the future, especially in industries driven by advanced technologies.

An excellent example of the value to a democracy of a much-better-educated public is given by the example of dissidents in the former Soviet Union. Andrei Sakharov was the best-known Soviet dissident, too prominent to totally suppress, but there were others. While these individuals tend to be among an especially well-educated small elite that all societies recognize are needed to compete in the modern world, some fraction of these people in any society is likely to penetrate a tyrannical system and eventually oppose it. This is often followed by an aroused public that will then demand progressive change. While this possibility may induce some reactionaries to suppress the creation of an intelligent group of potential dissenters, we can expect a large fraction of the more intelligent and successful leaders of modern industry in democratic countries to favor quality education in their own interest, even if they think it comes with a risk of dissent.

Failure to provide effective public education mortgages the future of any society today in many ways. History makes clear that people will cling to whatever seems to keep them afloat, physically and psychologically. This can involve moving forward in stages or even lurching violently backward. Ignorance often produces the latter reaction, until major human suffering finally awakens a people. Future chapters will return to this problem of ignorance, especially chapter 10, where education versus mere training is better defined and different kinds of ignorance are discussed.

ENERVATING SUPERSTITIONS

A person can be ignorant and not be superstitious. Or a well-educated person may find it difficult to overcome a particular superstition that has made a deep impression on her. Most of us have known examples of both kinds of people. Ignorance and superstition are not identical, even if they are easily confused and usually closely related. All of us are ignorant on some level and may be unaware of it, but some of us like to think we are not superstitious as the term is usually defined. *Superstition* can be defined as "a belief founded on irrational feelings, especially of fear, and marked by a trust in charms, omens, the supernatural, etc.; also any rite or practice inspired by such belief." A terse second definition from the same source simply defines *superstition* as "any unreasonable belief."[3] I adopt that definition in what follows. Note that this definition demands a *belief*. A person may have irrational *feelings*, as all people surely do, but to be superstitious the person must at least translate these feelings into some kind of belief.

The question asks what the value of superstitions is today, granting that some superstitions probably had value at one time from a Darwinian perspective, especially given their almost-universal presence in all traditional human cultures. Their value must have depended on the particular superstition and probably on the particular culture in which it took form. Perhaps the most obvious form of superstition that accords with the above definition is dogmatic religion, for even if we assert for the sake of argument that one religion may be true, all others that disagree with that religion's dogmas must necessarily be false. This argument applies strictly to the religions' dogmatic claims and not to the "functional value" these religions may provide their adherents, but it will suffice for purposes of this discussion. The Enlightenment was anticlerical primarily because of the power ecclesiastical authorities exercised in eighteenth-century society, especially over government and education, but most of the prominent Enlightenment thinkers regarded religion itself as one colossal superstition.

Not all superstitions are religious in origin, though many are. The source of the superstition that harm may come to those who walk under ladders seems hardly religious, though it may have gained force if some ancient cleric lost to history succumbed to an object dropped from a ladder while walking under it. The source of many superstitions is obscure, though we can speculate that most of them either offered useful advice for avoiding trouble or at least appeared to do so. Most of us would not deliberately walk under ladders that held workers carrying heavy objects, unless it was the only way to get to the other side.

Thus, beyond the question of the often-obscure origin of many superstitions is the more compelling one of the benefits and harm that holding them confers on the believer today. The rest of this section will consider that question.

Using dogmatic religion as a common example of a superstition (which does not in itself prove the falseness of the religion), we can say that the ratio of the beneficial to the harmful effects of a superstition depends on the particular superstition. If it could be shown by a rigorous, scientifically conducted study that a particular religious belief or ritual enhanced the physical and/or mental health of people in general, or even of certain definable groups of people such as sincere believers, then all reasonable people would be compelled to admit that the associated superstition had at least this particular functional value. An atheist who would scoff at this argument might recall that the strongest propensity built into our genes is not to seek the truth but to survive and reproduce, for which there is no better authority today than "new atheist" biologist Richard Dawkins. Even serious-minded secular philosophers will often refer to truth as an instrumental value, viewing truth as a recommended means to an end and not the ultimate goal of life itself. We have already noted this view in connection with the goals of the Enlightenment. The fact that most scientists take the concept of truth so seriously is highly pragmatic. Most scientists have become convinced by our reflections on life that in the long run nothing matters as much as the truth, because over time nothing contrary to the truth will work in practice.

Many other examples of the possible value of different superstitions can be given, but the above example makes the point. Chapter 7 reviews some of the ideas offered for the origin(s) of religion among early peoples, and many of these invoke the functional value of religious superstitions, especially for those early peoples who had no natural science as we know it today, and for whom a spirit world often made good sense in light of their limited reliable knowledge and their personal experiences.

So far, these comments may seem to suggest that superstition could confer more benefits than harm to people, even today. While not denying that this may be the case for some people today, it is certainly not true for all people and is, I think, not true for most people. This conclusion is based on considering the consequences of superstitious belief. A superstition sincerely believed to be true settles the question for the believer, requires no further thought, and if contrary to the truth may lead the superstitious person and others to engage in dangerous behaviors, either for the individual or even for the entire society. Some religious superstitions have an increasingly inhibiting effect on promoting human progress, because when proved harmful they can provoke a denial of reality that humanity can no longer afford to ignore. This is particularly disturbing when religious dogmas continue to promote a once-reasonable mandate that has become counterproductive for human welfare today. A good example is the still widely promoted religious teaching to avoid contraception and to continue to "be fruitful and multiply." This population policy, which is defended by reference to antiquated dogmas attributed to God, is in danger of populating the earth beyond the current carrying capacity of an attractive, healthy environment. A more rational approach is possible, but it may not be possible to apply this effectively as long as unreasonable beliefs persist.

Even in the absence of outright denial, superstition almost always deadens the mind to manifestly recognizable realities, enervating people by distracting them from real problems and often dissipating their creative energies in fruitless activities. Karl Marx is associated with the idea that religion is the opiate of the people. Because Marx is the author of this particular statement, which has been a common sentiment among many other thinkers through history, the reaction of many people is to ignore it. Unfortunately, this immediately reduces the number of people who are willing to examine the statement on its own merits, independent of other ideas from a thinker who advocated political solutions that most of us find both naive and dangerous.

When examined from the perspectives of history and anthropology, superstitions become increasingly understandable. History and anthropology give us some idea of how people lived in the past. Though there are exceptions, many scholarly studies support Hobbes's well-known comment that for most people living in the past, life was "nasty, brutish, and short."[4] There are atheists who argue it is a shame that earlier peoples did not adopt science when Thales of Miletus used empirical methods to predict a solar eclipse. These arguments are flawed by the failure to ask what was possible for most people living in 500 BCE and before. I would not attempt to make a case against the functional value of all superstitions throughout history.

To say that religion is the opiate of the people does not mean that many did not need a psychological boost to make their lives tolerable.

However, the relevant challenge today is to determine the consequences of superstition when, barring major disasters, most lives are no longer nasty, brutish, and short. What is the value of accepting literally the dogmas and urgings of "religions of the book" when weighed against the consequences of those beliefs, not only to society at large, but even to their individual adherents? How much comfort do superstitious people derive from embracing irrational feelings and adopting the resulting irrational beliefs, and how does this compare to the harmful effects of adopting such an irrational approach to resolving some of life's most difficult issues? At the very least, we can say that superstitious thought is mentally lazy, even if for some people it is the only possible way to achieve some inner harmony.

Superstitious thinking is akin to a mindless habit. We are all creatures of habit, and we derive many advantages from this. Learned responses to routine situations that are implemented with little thought save us time and energy for more important tasks. A superstitious person might ask us, "If we are aided by these unthinking habits as guides for our daily behavior, why not also accept comforting superstitions that make us feel good?" We might reply, "If this superstition assuages your guilt and justifies your antipathy for your enemies, why not try astrology to see if it will improve your personal life?" And so on. Finally, lazy superstitious habits of mind habituate people to seeking leaders who can inspire their followers with visions of victory over their enemies and solace over their personal shortcomings. Superstitious thinking of any kind almost always involves a denial of unpleasant realities that are often in full view of the denier, who feels compelled to seek an emotionally satisfying, easy way out. That is hardly a formula for finding optimum solutions to great world problems or even small personal ones. Yet we must admit this is "all too human" and cannot be eliminated easily. Thus these remarks concern only the first necessary step toward solving any problem, which is recognizing the problem in the first place. In a future world based more and more on reliable scientific knowledge, superstition is sure to decline and be marginalized. Getting there is the problem, but recognizing that this will take time becomes part of any reassessment of the Enlightenment.

JUGGERNAUT TECHNOLOGY

Ignorance and superstition may lie at the heart of many of our contemporary societal problems, but our highly productive modern technologies are playing a role as well, especially when they defy rational control. Juggernaut is a cult name for the Hindu god Vishnu. It is also a vehicle famous in Hindu mythology for riding relentlessly over all those in its path. The truth is more pedestrian, as the vehicle is actually drawn slowly through sand by devotees during a religious procession, but frenzied followers of Vishnu have on occasion actually thrown themselves under its wheels in religious ecstasy when the Juggernaut is driven through the crowd.[5] In analogy with a relentless juggernaut, certain modern technologies have developed a life of their own, running partially out of control over increasing numbers of people who cannot escape their relentless course. We might say in those cases that technology has become the driver and rider, while humanity has become the horse.

It would be unfair to conclude that we have reached such a pass overall, as most people still absorb and benefit from the majority of science-driven modern technologies developed since the eighteenth century. Nevertheless, the pace of development makes it increasingly difficult to predict where technology will take us next, except to say we will arrive there faster. Decades back, Marshall McLuhan noticed some ominous features of modern television, a medium that presents information so quickly that a clever marketer—or propagandist—can transmit a subliminal message that only a well-trained skeptic can recognize.[6] The racing-car ads crafted for the testosterone-driven male immediately come to mind. "Buy this car and you will attract a beautiful woman." Perhaps. Freedom to advertise and the need for some risk taking are essential in a healthy free society. So are an educated skeptical mind and common sense.

The human race has clearly failed over the past two centuries to prevent many of the destructive consequences arising from large-scale modern science and the technologies it has birthed. The two World Wars of the past century, the vast resources expended for military preparation during that century, and the global insecurities arising from the Cold War immediately come to mind. While military preparations and actions by the major democracies yielded a far better outcome than an Axis or a Stalinist-style victory would have produced, the huge expenditure of lives and resources that might otherwise have been dedicated to developing a more productive civilization was an enormous if necessary waste. Gaining a perspective on the benefits and the harmful effects arising from modern technologies suggests that we may be currently struggling through what an anthropologist might call an advanced stage of technocratic barbarism, at least if seen from a more enlightened future vantage. While I think such a description of our times would be extreme, we cannot claim that modern technologies have been only benign.

Another example of our failure to manage the benefits of modern technology without suppressing invention and becoming Luddites is our current inability in the United States to deal rationally with human-driven climate change, a serious global problem already noted here. The climate-change problem will be technically difficult to solve, even when we overcome the current lack of political will to solve it. Certainly, a more general recognition of the problem would be a healthy beginning, but well-funded interests have so far defeated efforts to educate the public on the problem's reality and the part played by human activities in causing it.

There are other global problems that should command attention today, where progress toward solutions is painfully slow. Nevertheless, next to preventing another large-scale global war with modern weapons of mass destruction, maintaining a healthy ecosystem for life on earth may loom the largest, with managing climate change a large part of that picture. Only widespread education on the severity of these problems seems likely to overcome the present resistance to change, partly due to political ads that in some cases might better be called propaganda.[7]

A more subtle question that is difficult to answer is what benefits and harmful effects are likely to come from modern electronic communication devices. There are many major benefits from a higher speed of information transfer. Other studies suggest that young people may think faster and better than their predecessors by using the latest handheld devices with capabilities that exceed those of main-frame computers of two and three decades past. These benefits would seem to exceed the drawbacks, but the impact on deeper, more conceptual thinking is less well understood. Technologists tend to favor developing a new technology free of all regulation and letting the problems associated with the new product reveal themselves along the way. This is probably often the only way to learn the many impacts of a complex new product, especially since technology assessment is a relatively new field that tends to be unpopular with many technologists, who regard it as stifling innovation. A technically savvy conservative politician effectively killed the US Congress Office of Technology Assessment in 1994 by simply not funding it. On the other hand, this same office made a remarkably accurate study of the real costs of NASA space-shuttle missions, which was embarrassingly high, demonstrating that technology assessment is a field that offers significant future promise.[8]

Much good work is under way in many areas to address these kinds of problems, and the prognosis for their solution is not entirely bleak. Nevertheless, the problem of regulating major juggernaut technologies remains largely unsolved. As in the case of controlled nuclear reactions, the prospect of runaway may be small, but the consequences can be devastating. For that reason, a preliminary discussion of some contributing factors is useful. By the definition I use here, a juggernaut technology is one that is running at least partially out of *anyone's* control. We can start by asking if there is a clear example of such a technology, one that plays a major part in our contemporary world and is in some sense being carried forward by forces not under the control of even its most skilled managers and technocrats.

Once more, the large global energy corporations offer an excellent example of what is both good and troublesome about major modern technologically driven businesses. These companies, their stockholders, and their customers all derive great benefits from the production of oil in an efficient manner, even while environmental costs due to their activities build up. These same companies and their investors also benefit from corporate laws enacted to protect stock holders. Current law requires management to emphasize profits for the investors, which can lead to liability suits against the managers. This invariably gives high priority to short-term profits, which could compromise judicious long-range planning for salutary alternatives requiring current investment that could lead to greater profits in the future. Assuming that the boards of directors of these companies and their CEOs decide to make major investments toward future operations, we can ask if these people could be in danger of losing their jobs, considering the possibility that short-term profits might slip.

The reader can see where this line of questioning is leading. The executives of these huge oil producers are not as free as they may appear to be. In one sense, they must be as wary of the juggernaut they do not quite control as the rest of us. Failure to maximize short-term profits and/or failure to support a successful lobbying effort to suppress climate-change legislation could lead to an executive turnover. The system itself works that way, and to assume it is driven by anything comparable in importance to "the bottom line" is surely naive. Most of the top oil executives and the best technocrats who work with them are very intelligent people. Finding and extracting oil is challenging and dangerous work, even with safeguards. These people are not ignorant, but even absent further government regulation, they are somewhat restricted in what they can do.

In effect, blaming all the wealthy executives of the current corporate world for its excesses is too easy and misses other parts of the problem. Those of us who are political progressives deplore the inequitable distribution of wealth in the United States today, where it is currently appalling, but demonizing one class does not help when the real problem is deeper and more complex. Chapter 2 briefly mentioned the familiar "structural problem" of insufficiently regulated mature capitalism. That problem is real, and part of the explanation of why many top executives who might wish to pursue a more enlightened course have their hands tied by the very system that brought them success in the first place.

All the issues raised in this chapter are revisited later, especially in chapter 12, where the interactions among the many pieces of this troubling puzzle are considered from a broad, somewhat speculative perspective.

PART 2

PROSPECTS FOR

PROGRESS

CHAPTER 4

THE CRITICAL ROLE
OF SCIENCE

SCIENCE, *THE* SUPERIOR WAY OF
UNDERSTANDING THE NATURAL WORLD

We have already noted in chapter 1 that science and reason were regarded as *the* best instruments for achieving the goals of the Enlightenment. This section will show why.

Science has such prestige today that even those uncomfortable with some of its results try to use it to legitimize their awkward unscientific positions. Examples of this range from so-called creation science to efforts to distort science to convince scientifically illiterate people there is either no global warming, no climate change, or, as a last resort, no human activity responsible for these changes. This passionate effort to deny selected parts of genuine science through bogus science is thought to be necessary by some, because of the many benefits that even the scientifically illiterate now know come directly from science. This is especially true when science fails to legitimize anachronistic views of human origins or criticizes contemporary technological practices that provide both important personal benefits to many and large profits to some if continued without changes today. Knowing that the great importance of science is grasped by a majority of people today, the merchants and propagandists of selective science denial have even been reduced to attempts at redefining science with terms like *sound science* to suggest that their "science" is real, while legitimate science is somehow false.

Real science can be easily identified as that which is subjected to peer review unless the research involves national security or company priorities, both of which can occur. However, since most of that research is of an applied nature, and since all science is based on research to understand the basic workings of the natural world, the peer-review process covers most of the fundamental work. This includes work in the biological sciences that has become somewhat controversial in certain segments of American society today. One should always be highly skeptical of claims by a scientist to a result that has not been subjected to peer review, unless the circumstances mentioned above apply, in which case some skepticism should probably still be retained.

It is equally easy to define what science is. Science is a process of investigating the natural world, sometimes called the *natural order*, according to a generally accepted procedure that involves (1) imagination, (2) observation and experimentation to collect data, and (3) rational thought, often expressed mathematically. When describing what is often called the *scientific method*, imagination is sometimes left out. That is a serious omission. Without imagination, science cannot move forward, since in the absence of imagination, the scientist is necessarily thinking within the system of what is already known. The most valuable research always involves new discoveries of what is not already known.

However, there is one important caveat to be added to this last statement, and it is fundamentally important to understanding why high trust can be accorded to the process of science, even though individual scientists, being human, can and do make mistakes. Obviously the best way for a scientist to make a reputation is to get it right. However, many fine, if somewhat lesser, reputations are also made by showing that a scientist has made a mistake, with all the positions in dispute appearing in the peer-reviewed literature.

The importance of these corrections cannot be emphasized too often. It says that science is a self-correcting process. No other profession can make that claim so emphatically. A subtle error may persist for a long time, but sooner or later, if the process studied falls under the purview of science (as the entire natural order does in principle), the truth will out. A very simple statement that the reader will read elsewhere can thus be made. As a process for understanding the natural order, science works. Failure to recognize this self-correcting process has led to some ridiculous statements made in the name of postmodern and deconstructionist thought, when certain authors unfamiliar with science try to justify the claim that all ways of knowing are arbitrary, suggesting that they are all equally valid. That statement, when applied to the workings of the natural order, is nonsense.

Before moving on to a demonstration of the superiority of science for understanding the natural order, it is useful to note what science is not. *Science is not a collection of facts about the workings of the natural order. Science is the process whereby these facts and the natural laws governing them are established.* If all that science has established to date were lost to us and if our descendants survive, science can recover and ultimately move beyond what we know today. If science itself were somehow lost and not rediscovered, that would be impossible.

From this I will now show why science is more successful, and usually much more successful, than any other way of knowing about the natural order.

To do this, I introduce two phrases, calling them (1) the way of science and (2) the way of faith. Place them at opposite ends of a sliding scale. The way of science is defined above. The way of faith is defined as that which is accepted as true on authority alone, with no verifiable evidence applied. Thus the way of faith becomes what many people call "blind faith." If a truth claim arises applying the way of science, it belongs on the science end of the scale. If it is embraced as coming from an authority one trusts alone, place it at the other end. If the claim can be justified in part by applying the way of science, but also requires a "leap of faith" for its dogmatic acceptance, put it somewhere between the two ends of this spectrum. For purposes of the following argument, it is not critical where it is placed as long as it is not at either end. The following argument will show why the way of science is a more reliable way of knowing the natural order than any other combination.

The argument begins by defining what is meant by "knowing." I define *knowing* as being able to assert with confidence the probable truth of a proposition, one whose truth is not logically implicit in the proposition itself, which would make it trivial. We have already defined what science is, but more can be said about the three elements already described: observation and experimentation, reason, and imagination. Science depends on the quantification of observation and experimentation, which is careful *measurement*. Reason depends on *critical thinking*, which is much more than logic alone. Critical thinking uses logic to reach conclusions but goes beyond mere formal logic to examine and critically assess basic assumptions. Both of these further definitions are important. The quantification of careful observations through precise measurement can remove all ambiguity up to the level of quantum uncertainty. The astute examination of basic assumptions not confirmed by precise measurements greatly reduces the probability that magnificent formal logic will lead to nonsensical conclusions. There is a well-known phrase in the world of computers that describes this perfectly: "Garbage in; garbage out."

Finally, we come to the most crucial element in scientific advance, imagination. The extraordinary insights of Einstein when developing special relativity offer an excellent example.[1] Yet we also know that our imagination can generate absurd fantasies. It is the unique advantage of science that it gives us a way to test our nontrivial speculations about the natural world, to see which ones prove reproducibly valid and which others are products of our imagination alone. No other thought process can make that claim.

It is now possible to proceed with the proof that science is *the* superior way of knowing about our natural order. Of course if this natural order is the only order of substantive reality, as opposed to a proposed supernatural order, then science becomes in principle the superior way of knowing everything important, including the working of our own minds and the cultures we create, *assuming that this kind of investigation is possible in practice*. That is a difficult question we cannot answer today, but advances in science promise significant answers in the future.

The proof proceeds as follows. We consider all other ways of knowing that are not science alone, then argue why science is the most reliable. To do this comprehensively (extensionally) would clearly exceed the capacity of a huge scholarly tome, for someone could always propose a new way of knowing that differs in some detail from the others we had considered. However, there is a simple logical shortcut that permits us to proceed and cover all these possibilities, however numerous.

To take this shortcut we need a further discussion of the way of faith, which lies at the other end of our sliding scale. I expand our definition of the way of faith as a way of knowing by adding that it partakes of the following two different characteristics: (1) *Personal experience* of anything that cannot pass the tests of a scientific demonstration (at least to date), which is then accepted as proof of a substantive reality outside of the believer's imagination. Many claims for the reality of God are of this kind. (2) *Trust in an external authority* for the truth of a proposition, which is then accepted by those who do not insist on further evidence or reasons. The authority is often a religious or political leader, but anyone accepted as such an authority will suffice for our purposes here. Granting that many assessments of important probabilities cannot be established by tests as rigorous as those demanded by science, it remains the case that element (2) is often offered by those whose trust is extremely uncritical.

Having now established what we mean by the way of science and the way of faith, the argument continues by showing that all truth claims arise from one or the other way of knowing, *or from some combination of the two.* Demonstrating that the way of science is superior to the way of faith for understanding our natural order then completes the proof that science is *the* superior way of knowing the natural order. That demonstration follows.

Start with the definition of *science* that appeared earlier in this section. To date, science has never failed to produce a deeper understanding of the natural phenomena to which it has been consistently applied. No other kind of human activity can make that statement. Numerous convincing examples of this are provided in *A History of the Warfare of Science with Theology in Christendom,* by Andrew D. White, a book that appeared in 1895 but still has resonance today, written by a former US ambassador and the person most responsible for founding Cornell University.[2] Anyone reading that book cannot fail to appreciate how science demolished many assertions of traditional religion in its former claim to precedence in explaining the working of the natural order, which all but the most reactionary of the religious have accepted today.

The second statement supporting the superiority of science makes the point demonstrated by Andrew D. White explicit. When the way of science has been applied to *testable* propositions advanced by the way of faith, it has demonstrated many of these propositions to be wrong. The best example for contemporary America is the pathetic attempts of the reactionary religious to defend the various doctrines of creationism against the well-established evolutionary biology that gives the correct general physical picture for the origin of the contemporary human being. An insightful account of a famous recent trial that settled that in favor of the modern scientific synthesis is Lauri Lebo's book, *The Devil in Dover*.[3] The creationists were caught in so many rationalizations and contradictions that the judge threatened some of them with contempt of court.

An especially tragic example of ecclesiastical negligence appears in the previously referenced book by White, who reports that during the eighteenth century, authorities in the realm governed by Venice stored a large quantity of gunpowder in a church. Religious authorities had asserted that ringing bells blessed by the Catholic Church would protect them from violent storms then attributed to witches, an assertion apparently believed by many ignorant people. A storm came; the bells were rung; lightning struck, igniting the gunpowder; and the center of the town was demolished with great loss of life. Soon thereafter, the Catholic Church decreed that all churches in the Italian states could erect "Dr. Franklin's" lightning rods. We must note a great change since in that venerable institution's views on science, but they were a long time coming.

The third statement in support of the superiority of science as a way of knowing our natural order is that there is no reliable scientific way of testing many of the propositions offered by the way of faith. Furthermore, no person has yet discovered a fragment of evidence that has passed the tests of science for any linkage between our natural order and an alleged supernatural one. On this basis, we can conclude that all claims made to date for anything supernatural are either untestable by science or have failed those tests when the tests have been made.

This situation is recognized and accepted today by some distinguished theologians, though not by all, especially when we go back in time. The Catholic Saint Augustine was a brilliant and subtle thinker, and was admirably human in many respects. Most of us are still amused and impressed by his plea "Lord make me celibate, but not yet." Perhaps this distinguished theologian recognized long before modern science that there was no secular way to prove the existence of deity. At any rate, Augustine made blind faith a virtue by asserting its superiority to reliable (let us say scientific) knowledge as a way to reach God. The devout Danish Lutheran Søren Kierkegaard made a slightly different argument that also depends entirely on blind faith. Famous for the phrase "a leap of faith," Kierkegaard admits with admirable honesty that he prefers belief to nonbelief and so will simply be a good Christian by taking this "leap" with no further attempt at proof. While secular people like myself find this a disturbing position fraught with risk in today's complex, contentious world, it is a position that a person can still choose today, and many do. Kierkegaard was no emotional fool. His knowledge of philosophy was extensive, and he is sometimes credited with being "the father of existentialism." Nevertheless, the third statement offered to demonstrate the superiority of science remains true and is extremely powerful, particularly when we consider the many brilliant, passionate thinkers who have sought empirical confirmation for their faiths throughout recorded history.

Assembling the above three statements, I think people who might previously have developed a contrary position would find it difficult to deny that the way of science is a superior way to know our natural order when compared to the way of faith. This remains true even when compared to any combination of science and faith. Since all positions that are not reached exclusively through science must necessarily depend at least partly on faith (reconsider the sliding scale and how the two ways of knowing cover all options for knowing), this completes the proof. Science is *the* superior way of knowing our natural order. To date, there are no credible competitors.

Two more supporting comments can be made. The first concerns the axiomatic nature of logical systems. Unless you subscribe to the notion of "absolute truth," which is not popular in most scientific circles, all logical systems are based on fundamental assumptions, or axioms, that are accepted because they appear self-evident. However, even seemingly self-evident statements may eventually fall to further investigation. There was general acceptance that parallel lines would never intersect, a proposition fundamental to Euclidian geometry that no less a thinker than Immanuel Kant accepted as an "a priori truth." Today we know this depends on the kind of geometry we are dealing with, so a more fundamental set of axioms is needed.

Some theologians and others like to argue that science itself is just as married to nonrational (not necessarily irrational) beliefs as religion, by saying that adherence to the scientific method is itself a kind of proposition, or "axiom," accepted without further proof, just as the assumption of a deity can be accepted. While this may seem to be a stumper, it is not. Scientists *do* indeed think the scientific method is the right way to approach understanding our natural order, but they *never* accept a nontrivial statement about that order without applying science to it, and so far that approach has never failed to date. Scientists have discovered from applying this method untold thousands of times the world over that it works perfectly in achieving exactly what is claimed for it. This differs dramatically from many claims made through the way of faith, nor can any other method make this claim just made for science. To this we can add that if our natural order is the *only* order that exists outside of the human imagination, then the human imagination, too, is a product of this natural order and can also be studied most productively by science, even if science is admittedly taking only "baby steps" in that direction now. This subject is considered in greater detail in chapter 5.

Finally, I am not saying that people should never accept anything on the basis of some degree of faith, although the word *hope* might be a more appropriate term. We would not travel on public conveyances if we had no confidence in the manufacturer or the driver/pilot. The rational person will often do a quick mental calculation of the odds of surviving under unusual circumstances but will also rightly assume that under normal circumstances things will probably work out well. I made such a calculation a few days after the infamous destruction of the Twin Towers in New York on 9/11. Several of us in NASA were scheduled to attend a science meeting on the Sorrento Peninsula outside Naples. The agency advised all scheduled NASA foreign travelers to consider not flying abroad until further notice. Several of us quickly estimated that the odds of some Islamic jihadist blowing up the particular plane we were scheduled to fly on were almost vanishingly small. It was a marvelous trip, and our Italian hosts were much more appreciative than the circumstances required. (Who is going to turn down the chance to visit one of the most beautiful places in the Mediterranean?) Many good things in life are worth a small risk, or life could become dreadfully boring. Evolutionary biology has taught us that we have evolved to be problem solvers and need to accept reasonable risks, to experiment, and to explore. We now know that progress and our descendants' survival depend on it.

That concludes this section. The next section proposes and defends a thesis concerning the future of our descendants. Given that many of the comments made so far in previous chapters may have seemed gloomy regarding our immediate future prospects—as some indeed are—it is a pleasure to say that what follows in the next section is quite optimistic, provided the reader is willing to take the long view that stretches well beyond any likely current human life span. Establishing the superiority of science as a way of knowing is an essential first step to that thesis, which is itself a vindication of the original Enlightenment approach to the future.

SCIENCE, HUMAN NATURE, AND A BETTER FUTURE

The original Enlightenment envisioned a world in which the opportunity for a fulfilling life would be offered to people everywhere, and the means would be available to realize it. Science and reason were recommended as the best way to achieve Enlightenment goals. This high regard for science is reinforced by the previous comments about science as the superior way of learning about the natural world. Today, even the great majority of the religious will acknowledge the usefulness of scientific knowledge, notwithstanding a minority who insist on the primacy of ancient texts when they differ from the modern scientific consensus.

However, this does not mean that science *alone* is capable of showing us the way to this better world. Knowledge may be power, but power can and often is misused for purposes that can hardly be called humanistic. The use of knowledge for humanistic purposes may benefit from certain kinds of knowledge, but the use of knowledge for benefiting people clearly needs to be guided by something beyond factual knowledge itself. Traditional religion agrees with this statement but then adds to it that this guiding hand must be the hand of God. Thus, the claim is often made by the traditionally religious that we cannot be moral without God. To this is sometimes added the famous sentiment of Ivan Karamazov (from Fyodor Dostoyevsky's best-known novel) that without God all things (including all crimes) become possible.[4] This claim has been accepted by large numbers of people, including some who are highly intelligent and well educated, but it needs to be challenged.

This section will argue that there are two attributes that humans share, which taken together can provide humanity with moral guidance. Both are entirely secular and require no gods. The first of these is our genetically influenced human nature. There are propensities built into our shared genetic makeup that prompt us to behave well toward others. The other attribute derives from the first and is the development of ethics. The final section of this chapter will consider the biological basis for ethics, anthropological evidence for how ethical principles have developed, and the relationship between ethics and science.

Given that moral behavior does not necessarily rest on a deity, it becomes possible to state the thesis upon which the long-term optimism for realizing essential Enlightenment goals rests. I assert that, despite all present and future difficulties we and our descendants will encounter, the odds for human survival and enjoyment of a higher state of civilization in the future are excellent. This thesis was stated in the preface and introduction and is repeated here with only minor elaboration. *The human race is almost certain to survive and develop further, both culturally and biologically. This will lead to the development of a higher level of civilization in which a large majority of future humankind will enjoy more fulfilling lives than today. The basis for this long-term view is twofold: (1) our biologically evolved propensity to live, to thrive, to socialize, and to think, and (2) the knowledge natural science has provided that will increase dramatically with time.*

I argue here there that there is nothing fundamental aside from the still-evolving human nature and constantly expanding scientific knowledge that can move humanity forward toward the more satisfying future world first systematically envisioned by Enlightenment thinkers. Stated in different words, our innate tendencies to live, to reproduce, to experiment, and to explore, all guided by other innate capacities for empathy and intelligence, along with an increasing scientific knowledge are what will carry humankind to a more satisfying future world. I further assert that all forms of superstition, which have tried and largely failed to do so, never can. We can credit superstition for having made major positive contributions to the growth of civilization to date, but it is impossible for me and many others to see superstition as more than a possibly still-necessary but declining participant in moving us forward beyond the present.

Worse, when superstition becomes reactionary and turns backward in time, as some movements in all major world religions have done today, without denying sympathy for those so entrapped, many of us regard this trend as the greatest contemporary clear and present danger to the forward progress of the human race. Ugly notions such as original sin that can be partially understood on the grounds of evolutionary biology and culture notwithstanding, human beings emerge looking much better when allowance for the challenges humanity has faced is made. In addition, science demonstrably works in providing us reliable, hence useful, knowledge of the world we actually occupy, while for the same purpose superstition does not.

Having stated this very pro-humanity and pro-science thesis, it now becomes necessary to recognize and examine the troublesome side to our genetic heritage. A frequently accepted position was expressed by Sigmund Freud in *Civilization and Its Discontents*.[5] There, Freud asserts that since our biological nature evolved over time under more primitive conditions, we retain natural propensities that do not always sit well with the requirements of modern civilization. Thus human fallibility is seen to be partly a product of conflicts between our biological predispositions and the requirements of civilization, leading to, or even requiring, some degree of either suppression or repression, or both. This vulnerability has undoubtedly influenced the emergence of the previously mentioned, highly antihumanistic doctrine of original sin in Western religions, with more subtle counterparts in Eastern faiths. While an unpopular notion among liberal thinkers (including many Marxists) dedicated to social engineering to advance civilization toward human perfectibility, Freud's thesis must be taken seriously, for there is evidence to support it.[6] However, modern nondogmatic theology could still replace the bleak doctrine of original sin with the reality of human vulnerability produced by the rise from barbarism to the demands of civilization. It would be far more humanistic, and there seems to be a trend in that direction, except among religious reactionaries.

The frequent resort to physical violence more common among men than women and now known to be enhanced by greater levels of testosterone found in typical men is a serious problem that cannot be attributed to culture alone. Likewise with human sexuality. The usual kind of sexual activity is still essential for human reproduction and is likely to remain so for some time, and sexual pleasure also confers many other obvious benefits on people, not the least being the pair bonding that makes stable families possible for raising the next generation. Yet even here there have been many problems that individuals and society manage only with difficulty.

Our genetic predisposition for self-assertion and reproduction have created many problems on all levels of human relationships, from the personal to the global. The challenge society faces in achieving a balance between our genetic propensities to assert ourselves and reproduce in a challenging natural and human environment, on one hand, and promoting the general welfare, on the other, is formidable. While this situation hardly justifies the dispiriting notions of sinful wretchedness that theologians may have once thought necessary to control human behavior under difficult circumstances, they do help us understand why these ideas may have emerged in the first place. While much of Freud's highly prescientific and speculative psychoanalytic theory has suffered from increasing critical examination, his notion of civilization and its discontents remains compelling.

In spite of many attempts to devise a satisfactory solution for people that will accommodate both natural tendencies and reasonable cultural imperatives, a case can be made that none are highly satisfactory to date. Part of the problem is undoubtedly that the "one size fits all" notions of traditional morality fail to accommodate the enormous variety of needs among people and cultures. Both traditional strict conservatism and free-wheeling, no-rules radical liberalism are easy to fault when applied to an entire society. Anthropologists have taught us that morals can vary greatly among cultures, but that all cultures have morals and need them.

Looking ahead, we clearly need further investigation of the balance between the benefits we derive from our genetic predispositions and the often-harmful contemporary effects deriving from those that were once adaptive under the more primitive conditions, during which much of our biological evolution took place. However, we can say that today we also have for the first time the possibility of bringing vital new knowledge to bear on this problem and our associated "discontents."

To date, attempts to address this problem have been almost entirely cultural. Proposed solutions have been offered by all the world's great religions and their many variants, by different schools of philosophy, and by political ideologies that range from the extremes of radical libertarianism to different varieties of hypothetically benevolent authoritarianism. While often-widely at variance with one another, all of these passionately crafted "solutions" have one thing in common. All are stoutly detested and resisted by many of the others. A universal in-depth solution based on culture does not exist.

However, there will soon be available to us a powerful new aid toward achieving better solutions to this problem, an aid that brings us back to science and arises directly from current developments in the biological sciences. For the first time in history, humanity is approaching the threshold of having the ability to modify the human genome. Thus biological human nature, which has been effectively fixed during our time on earth as a biological species, will eventually come under our ability to modify it. It is premature to judge on what basis this will be done and when the process will begin in earnest, but of one thing we can be certain. Barring unlikely catastrophe, *it will be done.* One possible goal of this development would be to create a more harmonious accommodation between our genetic propensities and our cultural requirements, both of which then become susceptible to coevolution.

I argued earlier in this section that the will to live and to thrive, aided by an expanding science, will ensure the survival and progress of our descendants. That argument can be expanded to add "in spite of barbaric expressions of some of our natural tendencies, which can eventually be corrected with further scientific knowledge." This says that the resulting change will be biological (phylogenetic) as well as cultural. This is still a very general claim and requires more for its justification. That the human will to live and thrive is strong is beyond dispute. A good example is provided by an excellent 2010 film, *127 Hours,* the true story of a lone young climber who cut off his arm that was hopelessly pinned following an accident, in order to survive. Beyond survival, our innate capacities for empathy and intelligence both need proper development, but that these *capacities* are built into our genes is equally certain. To these considerations we can add science as the way to better understand in more detail what these propensities are, and we can learn how to alter them biologically if that becomes desirable. Chapter 5 develops this idea in more detail.

As already noted, one goal would bring our biological selves into better harmony with our advancing civilization. Most experts and moral philosophers recognize that it would be foolish to attempt this with the limited knowledge we have today. One example of the dangers of abrupt action for human enhancement is the prospect of rapidly improving human intelligence without considering improvements to other important human faculties, like our innate capacity for empathy. The benefits of higher intelligence are obvious. Less obvious may be the dangers, if this is done foolishly, since we have survived to date by more than intelligence alone. Years ago, I participated in a discussion among several thoughtful people on the question "If you could do genetic engineering today, what major attribute would you improve first?" The majority, including me, voted for empathy. Several of us concluded that if intelligence alone were selected, humanity could conceivably destroy itself. Even if that is wrong, genetic enhancement must be done carefully. One possible guideline might be to prevent intelligence enhancement from surpassing humankind's innate capacity for empathy, which is the biological basis for compassion. Nevertheless, along with subtle changes in the human genome that are probably accelerating today with the current pace of environmental and cultural change, deliberate modification of the human genome is sure to occur as the knowledge to do so becomes available. It is thus imperative that careful development of appropriate ethical guidelines also be pursued seriously. Fortunately, this seems to be occurring now, hopefully by invoking the most reliable science.

This section has stated and developed a thesis that major attributes of our biological human nature combined with rapidly increasing scientific knowledge provide the basis for an optimistic assessment of our descendants' future. However, lest this treatment be seen as a crude form of "scientism," more must be said about the vital role of ethics in guiding this process. The final two sections of this chapter treat that topic.

A BRIEF DIGRESSION ON ETHICS

The subject of ethics has always been central to human activity. It is not surprising that ethics was central to the Enlightenment, which was motivated primarily by a universal concern for human welfare that was then expanded into the question of how this might be best achieved. While the application of science and reason was advocated to achieve the goal, the goal itself was based on the humanistic ethical principle that enhancing human welfare in the secular world should have the highest priority.

Thus, the fundamental importance of ethics continues, and it is surprising to see the subject mentioned so little in the popular media today, or in serious private conversations except when political adversaries are excoriated or villains are vigorously condemned. This raises the question of how much concern for what is right is a matter of what makes us feel good about ourselves versus a more sober assessment of an issue based on more careful reflection. While a good self-image has value and may reinforce the ultimate source of ethics and morality, our innate capacities for empathy and intelligence must also be applied to determine the best way to deal with an ethical issue.

Many people will use the term *morality* to describe what I mean here by *ethics*. I prefer to make a distinction, and will do so in what follows. In these pages, I use *ethics* to refer only to broad principles for guiding behavior, which are typically not broken down into specific instructions like "Do this!" or "Don't do that!" Moral codes tend to be categorical in this way, at least in how I will use that term here. There is a clear case for the value of moral codes, but there is also an obvious drawback. As moral codes become more detailed, it is increasingly likely that in order to obey one command it will become necessary to violate another. That can tie the would-be moral person into knots, and it often does. "Philosophy 101" offers many simple examples of this to encourage people to think. In one hypothetical case, you are hiding a Jewish person in your attic and the Gestapo knocks at the door and asks if this is true. Let's make it simple and say he asks this question to all homeowners, making it easier to say, "No." You are still violating a categorical imperative to always tell the truth. The distinguished philosopher Immanuel Kant wrestled earnestly with this basic moral question, particularly the question of whether one should always tell the truth.[7]

Serious dilemmas can arise in situations where people are conflicted between choosing an apparently moral action and protecting themselves more effectively from exploitation by others who are less ethical. Some of these scenarios have been studied in applications of what is now called *game theory*. While this theory can compute optimum win-win situations for all participants in the game if everyone is ethical, the theory has been further developed to offer optimum ways to punish cheaters, forcing them to be more "moral" in their own enlightened self-interest. This may involve punishing the cheater. When the ethical problem is multi-parametered, such complicated procedures may be necessary, but for the example given above, a sense of basic decency will usually suffice . . . if the person choosing has courage.

Not all morals are of equal value; some are more compelling than others. This tells us the best way to decide which moral values in a detailed code should take priority. Moral codes need to be seen as pyramidal hierarchies, not as equal-valued rules on a list. Rather than see a persecuted person murdered, the moral person tells a lie. If there is some suspicion he may be lying, the *courageous* moral person still lies (but no doubt carefully). To resolve this dilemma, rigid moral rules are not enough. Using ethics and morals as these terms are defined here, the hierarchy of moral rules that led to an ethical response was made possible only by resorting to something more basic than "morality," namely ethics, the realm of principles, which in turn is only made possible because we have an innate capacity for that *natural* attribute, empathy, without which all ethics and morality would be impossible. If people cared for no one but themselves, there could be no ethics or morality. Since people do in fact care for others beyond themselves because of innate qualities built into our genetic makeup, this tells us immediately that at least the proximate source of moral behavior is not a deity.

The importance of ethics will appear often in the rest of this book.

SCIENCE AND ETHICS

Most people realize that there is a difference between science and ethics, but some do not realize how much the application of ethical principles benefits from scientific knowledge. Thus we ask, *What does science offer ethics?* Science offers reliable knowledge that permits us to make more accurate assessments of the probable outcome of different possible courses of action. If the goal has been determined by the application of ethical principles, available scientific knowledge can often tell us what specific course of action will achieve that goal in the simplest way possible. The process will not work in cases still frequently encountered if the necessary knowledge is not available, but when it is, the process works very well. It is important to recall that we have first determined our goal to be ethical.

It is also important to be sure you have entered all relevant factors into your assessments. A simple if trivial example will illustrate this point.

Suppose a friend in Philadelphia has been suddenly stricken ill and you wish to reach Philadelphia as quickly as possible. Since you are a pilot, you discover by reviewing all options and doing simple calculations that flying your small private plane from Washington to Philadelphia and landing on the street in front of your friend's house would get you there most quickly. Unfortunately, in addition to breaking the law you could also cause a serious accident, thus violating the ethical principle of "do no unnecessary harm to others." Upon reconsideration, you discover that you have omitted one important factor in your calculation, that your mode of travel needs to be safe and legal.

On a more subtle level, science helps us understand why we are capable of ethical behavior in the first place. We have already noted why in general terms, and we will return to this question. The simple answer is that we have a natural capacity for ethical behavior that can be summarized by our innate capacity for empathy. However, how our minds/brains operate in detail to express this innate capacity is still a subject of research in several fields, some of which is reviewed in chapter 5. Gaining a better scientific understanding of these mental processes will shed further light on how to educate more ethical children, and on how to better manage and eventually reform criminal personalities. Finally, a much deeper knowledge of our mental processes may require a rethinking of our ethical norms, or at least of how we apply them. An example is the possibility suggested by certain experiments in neurophysiology that when we become consciously aware of making a decision, the associated neural events in our minds/brains have already taken place. Should these results stand and prove universally the case, this will require a serious rethinking of ethics and associated concepts such as personal responsibility, and of how society should best deal with irresponsible behavior.

An important proposition relating science and ethics is the naturalistic fallacy. This is the claim that it is not possible to derive a prescriptive system from a descriptive one. In the context of our discussion here, the naturalistic fallacy says we cannot derive ethics from scientific knowledge. It would be a bold claim to assert that we can derive "should" from "what" today, and most thoughtful people writing on this subject think this is not possible for a variety of reasons. But what about some hypothetical future state when the operation of our own minds/brains is understood to the point where we can say on the grounds of "hard science" that we know what the typical human being *really wants*, how this varies among individuals, and how—assuming equal progress in the social sciences—this can be realized in society? That such a state will ever be reached is problematic, but since science alone provides the most reliable knowledge of a natural order that excludes all supernatural elements, it becomes an interesting question to ask. The evolutionary biologist E. O. Wilson has asked this question in his book *Consilience*.[8] Wilson proposes that if we ever attain this level of detailed knowledge, we should be able to derive everything about our culture(s) from first principles; in effect, the reliable knowledge science has provided. Wilson is wise enough not to predict we will ever reach this state and concludes many of his chapters on a modest note, writing, "I could be wrong." That is a scientist speaking.

Speculation aside, we can say with confidence that the naturalistic fallacy certainly applies today. A more likely possibility for the foreseeable future is that the growing impact of science on ethics will continue. The difficulty of achieving consilience as described by Wilson in any time frame of current interest is staggering. Arguing that the naturalistic fallacy is trivial might best be considered a parlor game for intellectuals today. But perhaps not always.

Is science itself ethical, and if so, why? Most scientists and many nonscientists would say yes, but there may be no universal agreement on this. The answer depends on how much a person values the concept of "the truth." Science by definition and from its history is the search for the truth about how our natural order works, and few would deny that certain scientific truths are useful for promoting human welfare. Yet, we have already noted that in the Enlightenment tradition, science and reason (and by implication scientific truths) can be viewed as instrumental values, of somewhat lesser ethical weight than the overriding goal they serve for achieving a universal humanistic ethical goal.

In addition, while certain religious traditions value the concept of "absolute truth," most scientists remain skeptical of the concept, not often in principle where it may have value, but because from an operational point of view absolute truth seems unattainable. Finally, many people who are not scientists are far more interested in applications of science that provide them a better life than they are interested in science itself, which is understandable, if disappointing, to many scientists. Unfortunately, a subgroup of these people are so threatened by some results of science that they turn against it when it suggests that their ancient texts from a distant prescientific era err, and they argue that these texts should replace science where the two disagree. While that, too, can be understood, it is not acceptable to science.

Nevertheless, even people who distrust science could say with some justification that at least some of the science described in the next chapter does not support the claim that our deepest human yearning is to know the truth. Instead, these inherited biological propensities are more honestly interpreted as urging us to survive, reproduce, empathize, socialize, and thrive—and in some cases to vigorously attack and eliminate perceived threats. If scientific truths help us do these things, fine, according to this position. If not, some argue that we should embrace whatever works, even if it may ignore the truths of science. Where does that leave us?

I am aware of no compact logical proof for the excellence of science that would convince everyone. That some of us enjoy learning science and doing science is not enough. That some of us also have a love of the natural world and its many manifestations is also not enough. Many of us have a passion for science that draws us to it, not unlike the feeling that others of a different outlook have for traditional religious dogmas. Neither feeling is entirely rational. However, there is an important rational component that supports the scientific outlook. As long as we restrict our investigation to understanding the natural order, *science works and superstition does not*. To this we can add that in the absence of any supernatural order, science then becomes the most reliable way *in principle* of understanding *everything* in the order, including our most personal individual thoughts and feelings and the complexities of our cultures. That this level of understanding of ourselves eludes us now and may continue to elude us for the foreseeable future in no way weakens this statement.

A few years ago, I was giving a talk to a humanist group in Virginia, and I was asked what I thought was the proper attitude for a scientist *when doing science.* My answer came out too quickly and was not understood. "You should make no compromises, ask and give no quarter, and take no prisoners." The questioner was not impressed. "Sounds like Iwo Jima all over again," was his reply. I had to reassure him that this was an attitude not toward people but toward the problem *when doing science.* It may be true that science is one of the most *mentally* aggressive activities of which we are capable, but statistics show that scientists are also among the most law-abiding citizens of almost every society in which they are found. The exceptions almost invariably occur when a society is subjected to oppression, and then scientists are often found among the rebels.

There is some justification for this aggressive attitude toward problems and societal issues. It is likely that the only thing many scientists truly detest is ignorance, since science regards ignorance, not imaginary demons of the human mind, as the ultimate cause of human suffering, and reliable knowledge, including knowledge of that mind, as the best way to ultimately overcome it. That can inspire a strong determination to remove the veil of ignorance that separates us from understanding. It also provides no respect for claims of cultish or forbidden knowledge. So the best answer I can give for why many of us who work in science think science is ethical is because the evidence demonstrates that the reliable knowledge science gives us matters greatly today in solving many of our most pressing world problems, and it will likely matter even more for our descendants in the future. This is why most scientists are becoming increasingly disgusted with the attempts of certain groups to misrepresent science in the political realm.

With that, the following chapter offers arguments why the life sciences may have even more to offer humanity in this still-new twenty-first century than the rightly honored studies of the inorganic world and its laws . . . although scientists know the two cannot be entirely separated.

CHAPTER 5

SCIENCE FOR LIFE

Recall that the primary purpose of science as envisioned during the revolutionary period of the late eighteenth century was the improvement of human life. Science was also used to undermine superstition, in the name of which a powerful ecclesiastical establishment in Europe had enjoyed a strong grip on human thought. Liberating that thought was one of the primary goals of applying science and reason toward that end. Nevertheless, even in that effort the underlying goal was the improvement of people's secular lives. I assess the success of that effort further in later chapters, but there can be no doubt that most Enlightenment thinkers were convinced they were on the right path to human improvement.

PHYSICAL SCIENCE AND LIFE SCIENCE

All science shares the same general approach to establishing the workings of the natural world. Everything written in the previous chapter applies to both the physical sciences and the life sciences. Today, the division between the physical sciences and the life sciences is becoming increasingly artificial. The physical basis for life itself is becoming better understood, with fewer sharp distinctions between what is wholly "alive" and what resembles a life-form in some ways but not in others, especially on the primitive level of viruses, and in prion cells that appear to reproduce without DNA. Even on the level of extremely complex life-forms like human beings, the field of neurophysiology is raising some interesting—and to some people, disturbing—questions. Some of these questions impact on traditional notions of free will, and the similarities and differences between human mind/brain functioning and the manufactured networks in the field of artificial intelligence (AI). I will return to some of these questions later in this chapter. There is also a much less sophisticated general view of the difference between what is physical and what is clearly alive, and a rather sharp division between the corresponding sciences still persists in the popular mind.

This division is often very sharply defined when questions of funding research projects arise. While a huge generalization, most research funding that supports non-pharmaceutical industries and national defense tends to be for the physical sciences, while most funding granted for medical research supports investigations of biologically related problems. I would not deny the importance of all kinds of scientific research, and I would emphasize the dangers of directing the most basic research in all fields, especially by those ill-qualified to do so. That said, this chapter will suggest that, in this century, basic research in those biological sciences that bear on the mental functioning of the human being may prove to be the most revolutionary and the most important toward solving some of our most serious human problems of the current era.

This chapter will review briefly some recent developments in evolutionary biology, as well as in cognitive psychology and in neurophysiology, which is brain science in popular jargon. The review will be superficial from an academic point of view and will instead emphasize questions under investigation more than results so far achieved. That approach arises from my own lack of expertise in these fields, but it is at least partly justified by the fact that the fields in question are relatively new and are necessarily still taking small steps today, while taking them rapidly so that some tantalizing new results have already emerged. Nevertheless, what follows here will be tentative, for as new research results emerge, the picture of what the human being is really like will probably offer us surprises, some of which are almost sure to be initially controversial. Such has been the history of science to date, and it would be foolish to assume it will be fundamentally different in the future.

I will argue in what follows in this book that until a more reliable science of the relevant questions raised by these relatively new fields has been established, rapid progress in improving the human condition is likely to continue to be slow and sometimes painful. This statement applies to all manner of human relationships and to related issues of socioeconomic justice, democratic government, foreign relations, and the problems of war and other human conflicts. Even a reevaluation of ethics and the law may be required by certain results from the relatively new field of neuroethics. The traditional approach to all these areas has, of necessity, relied on a collective wisdom accumulated over millennia and transmitted to larger populations by both religion and educational institutions, of which the latter in the West were often the outgrowth of theological and monastic centers. While religion often claims primacy in the area of ethics, many anthropologists assert that the real origin of morality and ethics is human historical experience that is subsequently often codified and transmitted by religion. Most people simply do the best they can with what they know at a given time. Yet, as we gradually gain a more reliable, scientifically based understanding of how each of us thinks and feels about life, of how we remember (or fail to remember and deny), and of how we make decisions, it would be the height of folly to stick our heads in the sand and console ourselves with ancient superstitions. That would be the equivalent of asking a theologian to design an airplane that could actually fly, and fly safely. There was a reason why, in the early days of aeronautical engineering, the designer of a new aircraft was often expected to be in the plane for its first test flight.

EVOLUTIONARY BIOLOGY

For purposes of what follows, I will treat evolutionary biology as the natural extension of what Charles Darwin and Alfred Wallace launched in the nineteenth century, changing forever a more traditional picture of human origins and rooting us solidly in the natural order of things. Not all the problems have been solved, but an enormous number of important ones have been. As of this writing, an example of a still-unsolved problem is the origin of the first, most primitive life-form(s), but the origin of our species is now well understood. In addition, the Darwinian paradigm is strikingly effective in offering convincing explanations for many genetically influenced human behaviors and how they are realized in common attributes of different cultures. For those who cling to the now-discredited notion that there is no human nature, Steven Pinker, in an appendix to his book, *The Blank Slate*, has compiled a list of characteristics that all human societies share, despite a large range of historical and geographical challenges.[1] This would hardly be possible unless certain *propensities* were present in our common human gene pool. It is important to note that this is *not* biological determinism. Culture has enormous influence in directing these propensities over a wide variety of challenges people experience. Notwithstanding, the evolutionary paradigm is so powerful that Daniel C. Dennett has called it "royal acid," for even when expanded beyond biological evolution, it seems to dissolve many traditional concepts that it touches.[2]

The new synthesis of evolutionary biology developed largely by Theodosius Dobzhansky, Ernst Mayr, and others in the mid-twentieth century has held up remarkably well against all challenges of those wedded to traditional religious views of human origins and engaged in extreme rationalizations to deny it.[3] Combining Darwinian natural selection with the genetic insights of Austrian monk Gregor Mendel, this modern and now increasingly refined mathematical theory of how evolution proceeds has survived all scientific tests to which it has been subjected. The famous recent trial in Dover, Pennsylvania, mentioned in chapter 4, exposed the unsatisfactory arguments offered to support "creation science." Philosopher of science Barbara Forrest testified at that trial and helped to expose the weakness of the creationist position on evolution. Forrest, currently director of the National Center for Science Education, has continued to inform the American public of the absurdities of the creationist position. She notes from her experience teaching in Louisiana that many politicians seem unable to understand evolution.[4] Perhaps they are unwilling to embrace evolution for fear of losing office to a sadly uninformed electorate.

Nor is the problem restricted to the Deep South in the United States. The issue of whether medical researchers can use human embryonic stem cells (hESCs) for their research may be tied up in the courts for years, forcing investigators over the country to scale back their research by avoiding the most effective stem cells, which are easily obtained without "killing" a viable human fetus if the issue is viewed scientifically.

NEUROPHYSIOLOGY AND COGNITIVE PSYCHOLOGY

From the perspective of evolutionary biology, the human brain and associated nervous network can be viewed as the most complex and sophisticated of all the physical features that define the human being. This has led to evolutionary psychology, a still-controversial field in certain aspects, but one that offers the possibility of uncovering genetically influenced mainsprings for many categories of human thought, including cognition. Until recently, it was difficult to begin serious scientific research on the detailed operation of this extremely complex system of neurons, axons, synapses, and nerves, along with many associated hormonal and other electrochemical processes. The field of physiology has made impressive progress in some areas, and certain electrochemical processes associated with muscle motor activity have been studied successfully, yielding impressive results. Nevertheless, not surprisingly, questions involving the nature and processes associated with memory, consciousness, identity, and decision making have until recently defied serious detailed investigation and have been the subject of endless speculation, if not speculative mythmaking. Even distinguished philosophers of science with an entirely naturalistic perspective may occasionally offer arguably exaggerated claims for how well we understand certain details of our mental functioning.[5]

Who and what we are and how we behave reside in these elusive neural networks. For that reason, it is hard to imagine anything not yet known that could be more important to our collective future than understanding them better.

Neurophysiology is likely to prove even far more difficult than evolutionary biology to transform into a mature science with detailed predictive power. Neurophysiology carries evolutionary biology and physiology into the realm of the mind, which in the absence of some elusive and never-found spiritual factor in mental processes is the working of our physical brains and nervous systems. The human brain alone is comprised of tens of billions of individual cells that typically interact with one another rapidly in highly nonlinear ways. While some mental processes are largely restricted to well-defined areas of the brain, the once-popular idea of the relatively simple three-tiered "Broca's brain" that could be probed for easily understood correlations between electroencephalography and different observable behaviors has been generally replaced by a much more complex model. Consequently, given the staggering complexity of human neural networks, the progress in this field has been much slower than in evolutionary biology.

Despite this, human ingenuity continues to develop clever experiments to evaluate significant features of mental development, even without using the latest electronic technologies. The phenomenon sometimes called *agency*, a tendency of the mind to attribute reality and powers to imagined sources, has been carefully studied in small children and has produced significant results. Researcher Jesse M. Bering reports on highly suggestive research on infants and how their sense of reality changes as they slowly mature.[6] A young child is asked if the mother is present when she has disappeared behind a screen, and the child returns an affirmative answer. The mother may have left the building, or not, but the idea of her disappearance is unacceptable to the child, and the imagination provides the desired answer. The same child is presented a visual drama in which a model alligator devours a toy mouse. When the child is asked if the mouse is dead, the reply is yes. However, when asked if the mouse misses its mother, the contradictory answer is also yes. A few years later, the same child gives realistic answers. The rational faculties have obviously started to develop.

However, this development may never go to completion. It has been suggested that a susceptibility to agency might have once conferred evolutionary advantages on those who are susceptible to it. The person who assumes that a noise in the forest could signal the approach of a dangerous animal might be more likely to survive than one who dismisses all such clues, assuming that the former person does not also succumb to irrational timidity too easily. Also, attributing agency to imagined sources is clearly involved when a person assumes that unseen spiritual entities are playing a role in influencing natural phenomena, a subject we return to in chapter 7, which deals with the evolution of religion.

Notwithstanding these successes, a reliable science of human mental activity that yields the predictive power associated with a mature science cannot be achieved until detailed measurements of our mental activity can be made while this activity is in progress. While this level of understanding may lie well into the future, impressive progress in neurophysiology has begun. The development of instruments that permit different physical signatures of mental activity to be recorded while the investigator utilizes independent means to record the conscious perceptions of the subjects is one example. This research bears directly on the philosophical (and theological) problem of free will. Possible societal implications are discussed more in the following section. However, philosopher Shaun Nichols gives an excellent summary of some of the methods experimental philosophers (who are not neurophysiologists but who are familiar with some of their current results) use to assess the possible psychological origins of how many people understand the concept of free will.[7] Not surprisingly, though this approach lacks the precision of a mature science, it reveals popular attitudes that are both contradictory and yet quite understandable. When asked if determinism in human decision making undermines responsibility *in theory*, most of those asked answered yes. When the same people were given a practical example, the majority answered no, reminiscent of the struggles philosopher Immanuel Kant underwent when forced by his own logic to compromise a bit and move from his magisterial pure reason to a more practical kind in an effort to say something important about ethics.

Beyond the question of the how the mind attributes agency to real or imagined sources and of how completely our decisions are determined, is the question of how we perceive and how we understand. The field of cognitive psychology that investigates these questions will obviously benefit greatly from a more detailed model of mind produced by neurophysiology. However, there is one result the field has already produced that has major consequences for the main thesis of this book. An important discovery about how we learn is that when we first comprehend anything, the actual structure of our brain changes.[8] Learning changes the brain, and as anyone who has tried to learn a challenging subject like physics can say, the process is not instantaneous. The implications of this are discussed in later chapters, especially chapters 7 and 8, which discuss religion and politics, but one major consequence is that there is a limit to how quickly people can be expected to change their beliefs and behaviors, both as individuals and collectively. Small wonder some philosophers of the Enlightenment were excessively optimistic about the pace of human progress, not unlike certain thinkers today who expect a logical proof (or disproof) of an emotion-laden proposition to convert even the most rational human mind overnight. Generally speaking, our minds do not work that way.

Work on mirror neurons has given us insights into our innate capacity for empathy, though as usual, it may be premature to speculate that the effects are always benign. An innate capacity to see aspects of ourselves in special others may provide a secular basis for attraction, affection, love, and mutual support, as well as strong group bonding with special others. However, the same neuroscience also suggests that the role of mirror neurons, like the neurochemical oxytocin, could also provoke aggressive action against those who oppose the ones with whom we identify . . . the "mother bear" phenomenon. It may be best to avoid blanket judgments until more definitive research has been done.[9] Not surprisingly, current research suggests that the bonding tendency associated with these neural processes seems to have been first strongly developed in women (and in other nursing species). That would correspond to the traditional idea that the female of these species, including ours, may be naturally more empathetic than the male. If true, that would lead to other questions. Might women as a group be even better qualified than men to assume certain roles in modern society, which men in patriarchal societies have assumed largely for themselves? Further research would shed light on such questions.

IS BRAIN SCIENCE RISKY?

Before launching into a strong advocacy for the vital importance of scientific research into human mental functioning, it is useful to consider some of the arguments that are likely to be used against it. These range from the fears of "creation scientists," who maintain scientifically discredited views, to the more serious arguments of conservative scholars who think that rapid development of these scientific ideas could undermine traditional concepts of human dignity, free will, and personal responsibility. There are also understandable fears that malicious mind-control techniques will benefit from this knowledge. Finally, there is even some concern that research in the related field of artificial intelligence (AI) might benefit from research on our "organic" neural networks, assisting AI to develop rapidly beyond the ability of human beings to control it. An extreme variant of this latter concern sees humanity supplanted by globally networked computers, a silicon-based dystopia proposed by the innovative computer specialist Ray Kurzweil.[10] Some of these concerns are reviewed here.

It is not necessary to further counter the more simplistic arguments of those who refuse to accept the current results of modern science. Others have done that very well. However, the possible impact on society of people disturbed by studies of the human mind/brain must be taken seriously. This is especially true in the United States, where our science education for a thoughtful minority is world-class but is often appallingly bad for the rest of the population.

We begin by noting the disbelief and anger that research into the question of how the brain makes decisions could provoke. This could become quite intense if the latest research reviewed in the previous section proves true in the most general case and traditional notions of free will must be abandoned. It often comes as a surprise to many people that there have probably been thinkers since thought was first recorded who have solved that problem for themselves, though few would argue they were ever in the majority. To many secular humanists, human dignity results from our highly evolved innate capacity for empathy—the obvious natural basis for ethical behavior—and our innate capacity for intelligence. If traditional notions of free will prove illusory under the scrutiny of neuroscience, we can ask what difference this will make in actual human behavior. It is likely that after the turbulent reaction among those habituated to think of themselves as free agents dies down, life will probably return largely to what it was before, since the daily existential problems humans have always faced will not have changed. This is exactly what scholarly studies of an almost unimaginable disaster, the Black Death, revealed, as we noted earlier.[11] Humans are remarkably adaptive, often especially so when shocked.

At the same time, we must grant that the concerns of conservative scholars are real, or at least they should not be ignored. Some of the thinkers of the eighteenth-century Enlightenment recommended that those who held radical views should reveal them carefully, based on the belief that society as a whole needed time to accept the new ideas. Contemporary American philosopher Daniel C. Dennett has expressed an understanding of this position today.[12] Dennett does not exhibit the same anger over all religion exhibited by some other nontheists who share his naturalistic epistemology.

Some rethinking of the basis for society's laws will likely result from this reevaluation of the human being. Yet even here, a good case can be made that the process should be far less difficult and disruptive than many would claim, except among those so wedded to traditional notions that they cannot make the adjustment. There are undoubtedly such people. The brilliant but reactionary Jesuit priest Napta in Thomas Mann's novel *The Magic Mountain* provides a model for such an unfortunate person, whose mental rigidity compels him to exit life "not with a whimper, but a bang."[13] Nevertheless, the same process for developing a society's laws recommended by John Rawls would need no change under this revised understanding of who we are as human beings.[14]

Perhaps the strongest argument for not proceeding too rapidly in habituating society to revised notions of our existential reality comes from neurophysiology itself. We noted above that the learning process is not instantaneous. Learning slows down with age, explaining why older people adapt to change more slowly than the young; but even in the young this process takes time, depending on the unfamiliarity and complexity of what is learned. We also know that many of the great historical thinkers recognized what modern psychologists and politicians of all ages have known well, that reason is often, as David Hume noted, the slave of the passions. The fact that science has proved something definitively means little to the person who is passionately devoted to a different view, for reasons that are transparently emotional. Religion is not the only activity to exhibit this tendency. Even some scientists may not be immune to rationalization when they present agenda-driven arguments to deny global warming.

Other concerns over the rapid development of brain science involve fears of malicious mind control or even the possibility that it could lead to the end of human life as we know it. I find these fears greatly exaggerated but not altogether lacking in substance, especially the former kind. The phenomenon of controlling other people's thoughts has always been with us. Psychologists and perceptive caregivers have long noted that infants practice it very skillfully shortly after birth, to get what they want from their parents. (Why are the young of most higher animals almost always pathetically cute?) As we mature, we develop many techniques for manipulating others to get what we want, often unconsciously, and professional marketers arrive at a high level of sophistication using techniques that often differ little from what in authoritarian politics is aptly called *propaganda.*[15]

Much contemporary American advertising for high-performance automobiles is clearly directed toward young or otherwise macho males. While these cars are often expensive, the message is one of power and greatly increased chances of "getting the girl." Within limits, people in a free society should arguably be permitted to take their chances and learn, since all societies need a certain fraction of risk takers for their overall progress. Still, we could probably rate the skill of professional drivers hired for some of these ads to speed on winding mountain roads well above that of the typical thrill-seeking amateur. I would not try this on the Grand Cornish in the South of France before practicing under safer conditions.

So much for marketing. What about the potential for misuse of increased knowledge of our mental functioning by malicious intelligence agencies, criminal elements, and business sharks? We can quickly review these risks in order. Contemporary Hollywood movies may do a better job of alerting people to the dangers of corrupt intelligence operations than infrequently read serious books, but there may be a tendency to exaggerate the dangers in film for box-office purposes. Malicious tendencies in the Central Intelligence Agency (CIA) are a favorite Hollywood theme. Anyone who honestly thinks a modern nation in a troubled world can function without a well-trained intelligence agency is naive, even if both competence and proper oversight by elected officials are badly needed and may sometimes be lacking. As an American citizen, I can only hope the CIA is as competent as its counterparts in today's France, Great Britain, Israel, and Russia, listed here in alphabetical order for convenience.

As for clearly criminal activity and its close associate, sharp business practices, these will probably be with us at least until we can better understand the mental mainsprings for these behaviors, a process that now clearly still involves considerable speculation. That speculation could be greatly reduced with a more profound understanding of the human mind/brain than we have today.

A final critique of an accelerated program of research in neurophysiology is offered by those who fear a possible negative impact on the nature of the human being, a concern that goes beyond the impact on traditional concepts of human dignity based on traditional notions of free will. Ray Kurzweil seeks to remain alive long enough to reach a "singularity" that, he claims, is likely to be reached when Moore's law makes computers so powerful that they will link up through a future Internet, take charge of the world, and render human beings superfluous. That would be a game changer! Kurzweil hopes to survive until the proposed singularity arrives, so he can download the digitized contents of his mind/brain into this network and achieve a kind of silicon-based immortality.[16] It is easy to just say "Good luck!" but Kurzweil has demonstrated a great talent for computer innovation, and, like other people now calling themselves transhumanists, he should not be dismissed out of hand. At the same time, it is hard to think that a transfer of a human organic neural network to a silicon-based one will be achieved anytime soon, since we hardly understand the organic one yet.

The human neural networks we are only beginning to understand are the product of a somewhat ruthlessly self-correcting evolutionary process that has proceeded with many large "bumps" over three and a half billion years, producing an outcome that was not predicable but that has certainly been well tested for survival. It is clear we still have much to learn before we can say we understand our own neural networks well as a mature science. How can we be sure that evolution has not optimized this mind/brain process through the possibilities offered by the organic world? All data streams that carry information suffer from error buildup. Do the error-correction schemes researched for the computers we build work in the same way as the ones in our own neural networks? Are they platform independent? These questions need to be investigated before serious attempts at any kind of in-depth mind/brain and AI linkage can be reasonably pursued. A transfer of the entire contents of one into the other suggests a science-fiction horror scenario already explored, in which the transfer of errors produces a decidedly unwanted outcome. Yet, some linkage of the two is probably inevitable. Who would not benefit from an implant that would permit instant downloading to his or her brain of all the information available on a subject of temporary interest, after which the organic "RAM" could simply be erased? Of course the implant would have to work reliably and not induce unwanted side effects, such as a brain tumor caused by overheating. This research will no doubt continue.

There is a practical way in which people could prevent a computer challenge to human supremacy in the foreseeable future. It is hard to believe that these networked computers can secretly scheme together to take over the world while we humans innocently sleep, though it is an interesting idea. There are many people paying attention to these modern developments, and some hint of a computer-network conspiracy might be hard to hide from them. Given our increasing dependence on these machines today, clever techniques for controlling the network(s) are clearly needed, without suppressing the opportunity for free expression, and some are already in place. However, there is another solution illustrated in Stanley Kubrick's 1968 film *2001: A Space Odyssey*. This solution depends on identifying and isolating the offending element(s). Then the power is simply turned off.

Future Internet security is a challenging problem, but it remains one in which the principles can be understood. Science is the key to managing these systems, as well as it was to building them in the first place.

For those interested in ethical issues that arise naturally when human enhancement is explored, a group of generally young academics at Oxford University have reviewed them in the university magazine.[17] Probably best known is Nick Bostrom, cofounder of the World Transhumanist Association (now named Humanity+). More moderate in outlook, but still hopeful, are Julian Savulescu, director of the university's Uehiro Centre for Practical Ethics; Susan Greenfield, professor of pharmacology and most senior among the four; and Roger Crisp, fellow in philosophy. The debate among the four reflects many of the current issues.

These issues are serious. Lighter moments are sometimes needed for balance. Those of us who have lived long enough to recall a day of no personal computers, and who curse our laptops for interrupting us when we are working on them, delight in imagining the frustration future robots will inflict on those younger than we. Imagine a brilliant computer/robotics professor from MIT who arrives home in 2050 and instructs the house robot to open the garage door and serve the prepared evening meal. Sorry, says the robot, I have decided to take over the house and you may not live here any longer. The professor backs the car down the driveway, drives off, and spends the night with her colleague. She will solve the problem with the robot in the morning. Pity the robot who dares challenge its inventor.

On second thought, why are we so worried? Perhaps we should fear ourselves more than our machines. On a more humanistic level, perhaps our ignorance is what we should fear and overcome. Especially the still-prevailing ignorance of ourselves.

BRAIN SCIENCE FOR A BETTER WORLD

Having addressed some of the critiques of "brain science" today, I think that this field could contribute more to human progress than any other contemporary field of knowledge. For that, we currently lack the detailed predictive power associated with a mature science, but thanks to modern electronic probing techniques, progress in this field is accelerating rapidly. The argument favoring the critical importance of this field is strikingly simple and is difficult to deny. Our mind/brain is the seat of what makes us human. There reside our memories, our cognitive faculties, the seat of our decisions, and our sense of self; that is, our personal sense of identity. The mind/brain interacts with the rest of our bodies in complex ways and with the natural world and the human part of that world in even more complex ways. While it would be foolish to say we have a reliable scientific understanding of these many complex processes today, we can at least recognize the importance of achieving this.

The criticality of understanding our mental functioning can be further illustrated by considering the challenges arising in human activities ranging from the highly personal to the complex societal. All cultures develop guidelines for managing societal organization, family structure, human sexuality, personal and collective assertiveness (and aggression), economic requirements, and political arrangements. Anthropologists have already studied much of this in detail, in individual cultures and across cultures, as already noted. A better science of the individual human being, the fundamental unit of all societies, would provide a deeper insight into the origin of these cultural guidelines, which must serve both the basic needs of the individual while also providing for that person's productive integration into society. This understanding would also permit a reexamination of current cultural guidelines, some of which may require modification. In the past, these cultural guidelines were usually the product of trial and error. Some of them are easy to understand on the grounds of common sense. Food is a very basic need. In the absence of food, experience has shown that most people quickly lose interest in everything else. However, many other needs and behaviors are more complex.

The full range of human sexual and assertive/aggressive behaviors are important examples. Few educated people today would claim that Freud developed a scientific theory of human sexuality, or that Erich Fromm developed a correspondingly scientific theory of human aggression, though the writings of these two pioneers sometimes suggest that they thought so.[18] The relationship between sex and the difficulty of achieving full rights for women is discussed in chapter 6. One aspect of aggression is considered in what follows.

A major problem of aggression in our densely populated world today is the risk that nuclear weapons may actually be used. Preventing this in a world of competing sovereign states obviously requires diplomacy of a high order on all sides. Upon reflection, it becomes clear what a difficult problem this is for professional diplomats, each of whom is generally expected to seek subtle advantages for his or her own side. As a thought experiment, imagine yourself representing the United States in the 1970s, negotiating this nuclear question with your counterpart in the former USSR. You know how he perceives his role, and he knows how you perceive yours. Suspicion of "the other" develops naturally under these conditions, and in the example given it may have been only the universal fear of "mutual assured destruction" that made progress in reducing nuclear arsenals possible. A better understanding of how our individual minds/brains work under high group expectations and stress might relieve some of each nation's pressure on diplomats to deliberately seek advantage and instead to seek "win-win" solutions of mutual advantage. To date, powerful nations have often sought superiority in preference to mutual advantage. When two nearly balanced sides seek to dominate, there is usually only one likely outcome.

Accepting the recognized risks of acquiring more profound knowledge in brain science because of the enormous benefits this knowledge can confer is so compelling, it is hard to think there are intelligent people who fear the outcome so greatly that they would stoutly oppose it. Yet there are such people who rise to influential positions in society, sometimes through the democratic process. We can grant that there are dangers to moving ahead too quickly, especially when tradition-minded people are not ready for such advances. The educational problem clearly looms large here. However, barring global catastrophe, this progress is almost certain to continue.

Failure to better understand the human being from a scientific perspective virtually guarantees continuation of the endless conflict between and among different religious and political ideologies that has dominated much of human history to date. Only one human development in history offers promise to overcome the ignorance these conflicts thrive on. That development is science. If the primary humanistic ethical vision of the Enlightenment—a better world for people everywhere—is to be realized, it is hard to argue that there is any other human development besides science that offers this promise. Those who say that the more traditional approaches have not been adequately tried are forced to offer weak arguments when we examine the history of those attempts. I think we now have a growing understanding of why those traditional, often ideologically based approaches have failed. An excellent place to proceed, in my view, is in scientific studies of the human mind. If you, the reader, have an intelligent child who exhibits talent for science, you might ask her to consider entering the field of neurophysiology.

That is a good transition to the next chapter, which deals with human rights.

CHAPTER 6

UNIVERSAL HUMAN RIGHTS

HUMAN RIGHTS AND THE ENLIGHTENMENT

U nderstanding the primary goal of the Enlightenment leads directly to a consideration of human rights. Though science and reason are advocated as the means for creating a better world for people, the goal of achieving that world is fundamentally ethical in nature. This "better world for people" is one in which individuals enjoy more satisfactory and fulfilling secular lives in the "here and now." That in turn requires us to ask what human rights, among other things, will produce this state for people on a global basis.

The question of human rights can be approached in the same way philosopher of jurisprudence John Rawls suggests we seek to answer the question of what laws a people should produce for themselves.[1] In both cases, we would start with the individual human being, asking each person what rights they would like to have for themselves. The process then proceeds through successively higher stages with increasingly larger groups refining what previous smaller groups have recommended. Finally, the process culminates in a universal agreement on the fundamental rights that have survived this process.

Of course, this is an idealization, and the actual development of human rights has varied greatly in different cultures during different historical periods. Every anthropologist and most historians would argue that human rights have necessarily reflected the actual physical conditions of life in a culture, and that different conditions necessarily lead to somewhat different emphases on prioritizing particular rights.[2] The dream of universality has never been realized except in philosophical or theological idealizations. Nevertheless, certain common features have arisen in the basic human rights most contemporary cultures now embrace. Obvious examples are the rights to life and personal property that are universal in all significant communities today. Yet the idealized procedure described by Rawls bears some actual relationship to the way in which human rights have evolved over historical times.

The United Nations issued a Declaration of Human Rights, although it lacks the authority to enforce it; the French Revolution produced a Declaration of the Rights of *Man*; and the American Declaration of Independence that preceded both begins with the famous statement that all *men* are created equal and proceeds to declare their right to life, liberty, and the pursuit of happiness as "inalienable." Even these words are still constantly debated. Women were then often excluded from political participation and denied other basic rights, as they still are in many cultures and communities today. No educated person thinks all people are created equal in many genetic endowments, and numerous other caveats certainly apply.

Despite these reservations, a growing support for human rights has swept over much of the world during the past two centuries. The emphasis on universal rights that characterized Enlightenment thought played a role in this. While violations of human rights are often overlooked when compelling emergencies distract people's attention, the world has generally reacted forcefully to defeat the perpetrators in many extreme cases. Apartheid in South Africa was stoutly condemned by most countries and has since ended. The caste system in India is still followed in some circles, but the Indian government has officially condemned it and has created programs to improve living conditions and provide educational and economic opportunities for former "untouchables." Among the more egregious violations, no government on earth today could officially condone human slavery and survive international condemnation if it did so.

Today, genocide provokes universal revulsion, especially after the existence of the Holocaust became generally known. Unfortunately, that has not always been true. Those who regard the Holocaust as unique in history overlook the well-established record that the attempted extermination of a particular "pariah" people has a long history, as any student of the Mongol conquests knows. The initial enemies of the Mongols were Turkish peoples living in central Asia. They were slaughtered in huge numbers if the cities they inhabited refused to open their gates to the Khan's armies.[3] Today, the Turkish people are most numerous in Asia Minor, not central Asia, largely as a result. Even the Bible suggests that extermination of a people was a feature of intertribal warfare in the Middle East of the biblical era. The number of those killed generally becomes larger as the present is approached, but earlier populations were smaller, and their available technologies for this gruesome business were more primitive. Humanity has witnessed extermination before. While there is reason for optimism on human rights today, vigilance and public pressure are clearly needed where rights are still being violated for millions of people. An interesting recent study of how and why we *may* be rising to a more civilized state with less proportionate killing is Steven Pinker's *The Better Angels of Our Nature.*[4] We can hope that Pinker is right.

Major advances in human rights in the past few centuries have led one author to conclude that the idea of human rights is historically relatively recent. Lynn Hunt titles her book *Inventing Human Rights* and argues that literature played a major role in this process.[5] Hunt notes the role of the English novel in creating sympathetic characters who begin a quiet rebellion against a widespread aristocratic view that "common" people seemed to be there for their service and their pleasure. Painting and opera offer further support for this idea. Francisco Goya was famous for expressing sympathy for the ordinary people of his native Spain, being among the first painters to depict ordinary working men and poor women in duress in his still highly aristocratic society.[6] Mozart's opera *The Marriage of Figaro* pits an aristocrat's dancing master and his clever wife-to-be, Susan, against the Count, who is scheming to seduce the young woman, a key servant in the palace. Informing Figaro of the Count's obvious intensions, Susan allows her future husband to think he has discovered this plan, leading to a delightful song in which Figaro sings, "You do the dancing, Count, I'll play the tune." The servants emerge successful, and the Count is chastened. In Beethoven's only opera, *Fidelio*, Leonore saves her lover Florestan from an intended murder at the hands of a depraved Spanish governor by disguising herself as a boy who becomes Florestan's jailer. The night the governor slips into the dungeon, it is he who becomes the victim. The opera ends with the liberated couple and the chorus singing "Hail to the Day; Hail to the Hour" of their liberation. Goya, Mozart, and Beethoven, along with many other creative personalities of the Enlightenment, were strongly influenced by its advocacy of human rights. It is also interesting to note that in the two operas mentioned above, the real heroes were heroines, or women.

It is likely that Lynn Hunt's thesis is an exaggeration of the role of the arts alone in promoting the dramatic changes in many Western nations that came about during the latter eighteenth century. Such changes invariably reflect a confluence of many diverse developments and are often better understood after the fact. Contemporary thinker Nassim Nicholas Taleb, in his reflections on dramatic—and sometimes destructive—changes in society, refers to them as "black swan events" because of their resistance to prediction before the fact.[7] In addition, the frequently encountered romantic aspects of the arts may work well for inspiring enthusiasm for human rights, but there is a general recognition that romanticism in politics often leads to disaster. "Political opera" has long been a favorite tool of authoritarian regimes, from life-and-death spectacles in the Roman coliseum during the classical period to the dramatic propaganda films Leni Riefenstahl made for the Nazi Party in the 1930s.[8] Like science in the service of destruction—usually *not* inspired by scientists—the arts can serve two masters, and the one is not benign. More is said on that in chapter 8, which deals with politics.

Concern for human rights covers a broad range of issues today, from a variety of isms to be opposed to advocacy for full acceptance of individuals whose sexual orientation departs from the exclusively heterosexual norm. A growing science-based understanding has developed with the view that human variety is naturally greater than was once thought true, and that there are reasons for this involving the diverse productive roles that different individuals can play in society. The traditional notion that "one size fits all" in the realm of personal ethics and morality is being gradually replaced by a more subtle, nuanced, and insightful approach. This new approach will benefit greatly from the growing scientific understanding of the human being described in the previous chapter. In some areas where the influence of hormones on human development has proceeded, this has already happened. This more enlightened view is now widely accepted by people to whom tolerance comes easily, though it continues to threaten those who are uncomfortable with others who differ from them.

It is impossible to cover all these issues here, even superficially. Partly for that reason and partly because there are two major human population groups still subject to serious discrimination, the remainder of this chapter will review the challenge of countering racism and overcoming sexism.

COUNTERING RACISM

Racism in America is usually thought of as a black-white problem, though it is interesting that while Americans of largely African origin are now called *African Americans*, there seems to be little enthusiasm for the term *Euro American*. It is hard not to see residual European white-centric racism in this neglect, especially when we also use the terms *Asian Americans* and *Mexican Americans*. It would seem that Euro Americans have unconsciously granted themselves primacy in the social pecking order, though this can hardly reflect primacy of historical settlement, which must be granted to Native Americans, who were actually the first Asian Americans.

Thus it may come as a surprise to Euro Americans to learn that most slaves in the classical period of Western civilization were not black but white. Europeans of that period, and near-Eastern peoples of that period and before, often enslaved captives of defeated enemies and used them for a variety of menial tasks, though the conditions under which slaves lived during the Roman Empire were often much better than those of most of the slaves during the much later period of plantation slavery in the

Western Hemisphere; and many Roman slaves were permitted to earn their freedom. Princeton historian Nell Irvin Painter has presented a fascinating account of how this version of white slavery developed (not the current version associated with prostitution) from the early classical period to the modern era. In her book *The History of White People*, she argues that since the time when white women slaves from the Caucasus Mountains became the ideal for beauty, the concept of "whiteness" (i.e., whiteness is good) has been continuously expanded.[9] Today, it has been extended to practically everyone, including, finally, African Americans, at least in the minds of the better educated. The election of the current American president, Barack Obama, seems to reflect this development, though there remains both overt and covert evidence for residual racism toward the man, who is half black and half white. Attitudes in societies often change slowly.

One can still find people in the United States who, in the security of their family conversations, will aver that blacks will never be able to compete with whites in matters of the mind. Those of us who trace part of our own family histories back to the Southern states in America probably had at least one forebearer living when we were young who thought this true, and these people were sometimes well educated in the nineteenth century. Indeed, in the Painter book mentioned above there are detailed accounts of the pseudoscience of phrenology that were popular as recently as the nineteenth century (phrenology is the "science" by which careful measurements of the human head were used as indices of evolutionary development and intelligence). The mountains of data accumulated bear stark witness to the need to think carefully about what to measure before devoting a lifetime to doing so, or to the popular truism that correlation is not cause, should any correlation be found in the first place. Perhaps Euro Americans are less enthusiastic about such measurements today, when evidence suggests that East Asians may have the largest brain volume of all. Since brain size in humans has never been shown to correlate with mental functioning, should anyone really care?

In *The Road to Disunion*—primarily an account of how America slid into the highly destructive Civil War—historian William W. Freehling describes how cleverly plantation slaves often managed their masters and established friendly relations with the plantation owner's wife, who frequently sought their advice on important personal matters.[10] This seems to have occurred often when a woman was seeking information on the suitability of certain young men for her daughter's marriage. The highly intelligent kitchen slave often had the best available information on a young man's suitability through her own network of informants, increasing the chances the "owner's" daughter could avoid marriage to a debt-ridden, womanizing, hard-drinking dueler.

With emancipation following the Civil War, many deals were struck between politicians and financial interests in the North and former Confederate generals to end animosities as quickly as possible, to reunite the country, and to proceed with American economic development. While in some respects this can be viewed as a positive development, it was generally not beneficial to the freed slaves and their descendants. It was thought by many prominent (white) individuals, North and South, that the best role for the former slaves was to remain an often unskilled labor force, and myths about their congenital unsuitability for other work abounded. When that did not work in suppressing descent, a more extreme method of control was invoked: lynching.

In *On Lynchings*, early civil-rights worker Ida B. Wells-Barnett, the daughter of slaves, describes her struggles to stop the practice.[11] Her personal crusade began in 1892 when three friends were murdered for defending their store against a white mob, resulting in one of the attackers being shot. Some attackers then broke into the jail cell where the store owners had been detained, dragged off the black defendants, and hanged them. Lynching became a common way to intimidate the entire black community, as it provided a public spectacle as well as humiliation for the victims. In the 1880s and the 1890s, over one hundred African Americans were lynched annually, with 161 cases in 1892 alone. This intimidation of African Americans continued well into the twentieth century, with the last public lynching occurring in 1936.

That lynching or the threat of it appealed to many Euro Americans is clear from the popularity of one of the best-known silent movies of the early part of the last century. D. W. Griffith's film *The Birth of a*

Nation, based on the novel *The Clansman,* extols the "virtue" of the Ku Klux Klan (KKK) in defending white women from the "animal sexuality of the predatory black male."[12] While the KKK is less visible today, it experienced a significant renaissance in the 1920s, when it was reorganized at Pine Mountain, Georgia, and gained some political influence in numerous states, most notably in a nominally northern state, Indiana.

Racism was not just a Southern phenomenon in America, nor was it historically—and it is not so today. Our admiration for the righteous actions of many New England abolitionists before the American Civil War must be tempered—or perhaps enhanced—by the realization that some were probably direct descendants of sea captains who transported many of the earliest slaves from Africa to Charleston. Abolition by state often occurred where slavery was no longer lucrative. There were always places farther south that would purchase the slaves where the institution was still profitable, so abolition could proceed with little or no financial loss to the former owners. As is typical in human affairs, righteousness must be examined from many perspectives.

Shortly after the Nat Turner slave rebellion in 1831, the legislature of the state of Virginia debated gradual manumission in that state, and the measure lost by a slender majority. Turner's rebellion seems to have ended the chances for passing that measure.[13] Nevertheless, Virginia's attempt at manumission raises the question of whether the American Civil War—being celebrated in 2011 with some fanfare—would have been as long and bloody if then prestigious and powerful Virginia had ceased to be a slave state. History often turns on small local events.

For two decades, author Norm Allen served as executive director of African Americans for Humanism at the Center for Inquiry in Amherst, New York. During his tenure there, he organized a major program that included a regular newsletter and the foundation of several associated secular humanist centers in Africa. Allen also published two books of essays written by both African American and African authors. Several of the essays and the interviews Allen conducted with four of the authors are collected in *African American Humanism.*[14] These essays demonstrate that many ideas educated Euro Americans recognize as features of Enlightenment thought are also known to thoughtful members of this other community, and not all of these ideas were entirely indigenous to the European West. The hope for a more just human society would appear to be universal. It is noteworthy that the theme of many of these essays is secular. Some of the authors also grant that traditional religion may have addressed the needs of many earlier black communities, but they argue that secular humanism affords them both better opportunities and more hope today. Religion in African American communities is thought by these authors to often create a personal dependency that inhibits the initiative needed to solve people's individual and communal problems.

Allen describes his own arrival to humanism in the final essay in *The Black Humanist Experience: An Alternative to Religion*, which includes contributions by other authors from both America and Africa.[15] Like the majority of Americans, Allen had some religious instruction as a child, but unlike many with parents dedicated to dogmatic religions of the book, he had tolerant parents. This permitted him to experiment with different religions, different political ideas, and philosophy. The result was a long intellectual journey to humanism, a journey available to anyone willing to make it, independent of all categories into which people are placed for the convenience of or the exploitation by others. There is only one requirement, that the individual be human and willing to think. The protean thinkers of the Enlightenment and their counterparts in all cultures understood this well, but the message "humanity is one" still needs more vocal advocates.

This section concludes by asking if the horrors of slavery and subsequent racism have been exaggerated in the American experience. The argument is sometimes heard that many slaves were treated very well, and became "family" to their owners. Certainly human beings vary in their empathy for others. It is likely that on a fraction of the antebellum plantations some semblance of this scenario may have been realized; though it is doubtful this was generally true, and there are documented cases of extreme brutality. Because slaves were a valuable economic commodity, it is not surprising that some attention would be given to their health and general welfare. An interesting and somewhat-surprising result emerged in a study conducted by Robert W. Fogel and Stanley L. Engerman in their carefully researched book *Time on the Cross: The Economics of American Negro Slavery*.[16] Using modern analytical techniques for assessing the historical accuracy of surviving Southern courthouse and plantation records, the authors were surprised to discover the relatively high life expectancy of slaves in the antebellum period. They discovered it was due largely to a surprisingly good diet, one rich in sweet potatoes. However, in agreement with author William Freehling, mentioned above, they found no evidence of a single slave, even the best treated and respected of the kitchen help, who, given the choice, would not gladly have exchanged slavery for all the challenges and uncertainties of freedom.

This review of racism in America has concentrated on the early and middle stages of race relations in this country. The last fifty years have seen enormous progress here and elsewhere. With an American president of mixed racial background a popular world figure today, this is one of the most encouraging markers of human societal progress since the bleak days of government-supported slavery.

Likewise, there is another group whose members have made huge strides, and who compose the largest easily identified group of humans globally. Women have advanced to some of the highest leadership positions available in many countries during the same fifty years in which African Americans have advanced rapidly. However, just as racism has not been completely eliminated, neither has sexism. The last section of this chapter reviews sexism and offers what to some might seem a surprising conclusion. Sexism could prove to be the last major ism to be overcome, and might also be the one that, when overcome, would offer the best hope for advancement to a higher stage of human civilization. I will argue that this conclusion arises from reflections on generally understood power relations, but also from the nature of human sexuality.

OVERCOMING SEXISM

Those who argue that America has overcome sexism are easily shown to be wrong. Sexism is very much alive as these words are written, here as elsewhere, and there is little reason to think it will be overcome soon. Particularly damaging are the efforts of some men in influential positions who argue that "traditional values" underlying limited rights for women must be preserved. Because these arguments uphold a patriarchal status quo, it is useful to identify the beneficiaries. When examined carefully, the beneficiaries often prove to be the very persons—usually but not always men—advocating these limitations. However, before changing a patriarchal basis for societal organization to a gender-neutral one, it is necessary to show that the latter would be better. Today, we have reason to think this is true. This section will consider that question.

Perhaps the two issues where the limitations on women are most obvious and egregious today are the questions of who controls the bodies of women in their reproductive roles, and why women on the average still often fail to receive equal pay for equal work. When the question of equal rights for women arises, the arguments opposing equality can take bizarre turns. We can note some of the more bizarre arguments used several decades ago to defeat the Equal Rights Amendment (ERA) to the Constitution that almost became law. Imagine the disaster that would result if we had unisex public bathrooms, even when provided with security cameras in public areas and stalls appropriate for privacy! (Of course, in many places today, we do have them.) On a more serious note, if the ERA had passed, it would be much easier to challenge all the limitations that reactionary clerics and tradition-minded judges attempt to impose on a woman's right to choice in reproductive matters or on court cases that address unequal pay issues. When an argument is made that women's rights are already covered by existing laws, these examples provide convincing evidence to the contrary.

Certainly fear that society will undergo fundamental changes that are difficult to predict is one concern social conservatives have for granting women equal rights. Not surprisingly, while unexpected consequences usually follow from major societal changes, those who favor women's rights think the benefits these will provide far outweigh the problems that might result. However, since what is often called patriarchal authoritarianism still lies at the roots of even democratic societies today, it is useful to first examine why this traditional mode of societal organization, albeit with great variety among different cultures, may have once been optimum for advancing civilization before the rise of modern science and the Industrial Revolution. By determining why patriarchy under former conditions may once have been optimum, a strong case for equal rights for women today can be made by showing that the earlier conditions that favored patriarchy no longer apply.

A scholarly study of this question is well beyond my knowledge, so I will try to describe only what is obvious, well known, and well documented. Perhaps most obvious and clearly of great importance during earlier stages of human societal development is the significantly superior upper-body strength of the average man compared with the average woman. Even today this factor cannot be ignored, as the many examples of male physical abuse of women contrast sharply with the relatively few examples of women physically beating men. Some cultures continue to condone this practice today, especially under conditions where traditional notions of "honor" are still enforced.[17] Given the need for some degree of social order to support all the complexities associated with rising civilization (a political system, divisions of labor, stability for raising the next generation, etc.), it is not surprising that the identifiable group possessing superior means of enforcement should emerge as the leaders in early society, when brute force played a much larger role in daily life than it does in more economically developed societies today.

Under those conditions, it is also not surprising that men would appropriate for themselves the roles of preceptors and general authorities on all important subjects, including a few where one might think a woman would have at least equally valid qualifications. The areas of human sexuality and infant care immediately come to mind, yet the authorities most often quoted still seem to be male. (That may be changing today, as more women become doctors, psychiatrists, and authors for books on infant care, replacing the Dr. Spocks of an older generation.) Finally, and perhaps of greatest weight in the former relegation of women to predominantly domestic roles, were the critical reproductive and nurturing duties they had to perform to ensure the group's collective survival. In earlier times, human populations were smaller and more vulnerable to natural disasters.

It is easy to grasp why early societies demanded a primarily reproductive and nurturing role for women, while the men took care of other compelling tasks such as government, religion, hard physical labor, and group defense. More prestige was generally accorded the latter activities, in view of their obvious importance.

Granting all this, it is still tempting to suggest that the level of civilization a culture has reached can be measured by how well human rights in general and women's rights in particular have fared. While a simple linear relationship would be a foolish oversimplification that ignored the actual challenges of a particular culture during a particular era, this generalization seems reasonable. For example, if large-scale warfare among sovereign nations should erupt again as it did less than a century ago, it is hard to imagine women, many of them mothers, submitting themselves without protest to the kind of mass slaughter exhibited by large male armies, even among women living in aggressor nations. To the extent that enhanced women's rights might correlate with a more peaceful world, the impact on civilization would certainly be positive.

These questions have not yet been settled. However, there are indications from relatively recent research that there may be subtle statistically significant differences on the neurophysiological level between men and women, taken as two separable groups for study.[18] These differences raise the question of the roles the *majority* of men and women might elect to play in society, if truly equal opportunities were offered. While this may seem to justify many traditional roles for women, and may be used by conservative thinkers to oppose equal rights for women, the traditionalists could be in for a surprise. Lawrence Summers, a former president of Harvard University, was apparently asked to vacate that position because of a statement he made concerning the suitability of women for high-level scientific research. The statement seems to have been made in light of certain research suggesting that women are less likely to perform well in certain scientific fields. Even if this proves true, there is a rapidly increasing number of excellent women astronomers now that this field has opened its doors to them; and a growing number of women have recently entered physics. However, if Professor Summers had added that some research suggests that women, given the opportunity, might be *even more attracted than men* to fields such as biology, anthropology, neuroscience, and the more applied areas of medicine and psychiatry, perhaps he might have kept his former position.

Granted, these are areas of great contemporary controversy, with few firmly established results. (Note in passing that this is another argument for enhanced research in the neurosciences, as argued in chapter 5.) Nor are all those who may challenge research that questions gender identity necessarily opposed to women's rights. There is a passionate strain in radical feminism that insists on the identity of all mental functioning between the two genders, despite scientific evidence to the contrary. Since this radical feminist position can be questioned by some of the science that has already been done, adopting this dogmatic stance could prove politically counterproductive to actually advancing women's rights today. In contrast, the ERA could hardly be called a radical measure, has support from many men, and could still be passed.

All research on the question of male-female differences remains highly controversial today, and much of it is still in an early stage of development. Like other scientific fields that affect people's economic interests, personal welfare, or uncompromising ideologies (climate change, reproductive rights, biological evolution, etc.), legitimate science on gender differences has to contend with pseudoscience and ignorance.[19] The question of where women may have a natural edge over men for certain endeavors, and its obverse, are examples of questions that are still difficult to answer definitively, and in some cases are almost "too hot to handle" socially and politically. Nevertheless, it is science alone that can ultimately answer these kinds of difficult questions, especially in our contemporary world of clashing rival ideologies that often ignore science or even deliberately distort it or oppose it.

One area in which further research would be useful is a better understanding of the specifically sexual relationship between men and women, much of which is now based on traditional wisdom gathered over earlier times and on prescientific psychological speculation. While the role of hormones, pheromones, and other factors is becoming better understood—and the obvious sexual functions have long been well understood—the in-depth psychology of human sexuality remains prescientific. It is easy to give an example that bears directly on the question of equal rights for women and on one major reason why achieving these rights may prove difficult.

The example concerns the nature of courtship for sexual relations, the biological impulses that activate the male, and those that stimulate a receptive response in the female. Courtship among other, nonhuman "higher" animals can be easily observed by those who study them in a state of nature. Many nonhuman animal species perform mating rituals that are undoubtedly based largely on impulses the participants do not understand but to which they readily respond. For example, the female bighorn sheep (ewe) will often challenge the male (ram) to a dangerous run over steep mountain trails to see if he is "worthy" of mating. The female redstart warbler will play coy while the male engages in an energetic display of colorful feathers. Most of us have seen such mating rituals on nature programs, and some of us have observed them in the wild.

With humans, the issue is more complex. We like to think we know what we are doing and why. However, we are also strongly and subtly influenced by genetics, along with the other animals. So it is not obvious that all these genetically influenced propensities are well understood beyond the realm of useful traditional wisdom and prescientific speculation, though we do have the advantage of being able to think about the matter. Biologist Richard Dawkins notes in *The Selfish Gene* that much of our behavior has so far eluded in-depth scientific understanding.[20] Dawkins calls the other major co-determinant of our behavior the *meme*, his term for powerful cultural concepts of which religion is a prime example. Whatever terms we use, it is the interaction between these two co-determinants that largely guides our behavior. Dawkins notes that with increasing knowledge the day should come when the meme will exceed the gene in influencing that behavior, but he is not convinced this condition has been reached yet. Genetic propensities that have evolved to promote our species' survival may remain partly inaccessible to detailed understanding even today. Given the early state of mind/brain science, that would not be surprising. Upon reflection, the relevance of this to women's rights should be obvious, and some still-speculative unanswered questions will illustrate why.

Will equality of the two sexes mainly enhance or actually diminish the sexual relationship? If the answer is enhancement, then the case for equality becomes stronger. Otherwise, in view of the importance of satisfactory sexual relationships for life, the reverse *could* be true. The traditional view states that the man must be, or at least must feel, dominant in sexual activity, and there is a corresponding view (Freud's, for one) that a woman is best served by "surrendering" to the man. Is this true, or is it merely a rationalization to support patriarchal dominance in all realms of life, often "sold" to women to maintain a male power structure? Until this question can be answered on a more reliable scientific basis than we have today, it is risky to conclude that we fully understand all that is important in sexual relationships between men and women. Furthermore, since gaining equal rights for women will be difficult to achieve without the support of a growing number of men, knowledge of how many men actually feel about women may prove crucial for knowing how best to reach this goal. If a large number of men are actually threatened by women for *any* reason, or lack sexual self-confidence and are too ignorant or too proud to admit it, the goal of achieving women's rights will remain a formidable challenge.

The relevance of this discussion to the question of how and when we can expect most people to support equal rights for women may be large. Even if we make the case that appears in the following section, that equal rights for women would benefit the entire human race and contribute significantly to solving several major world problems, this may be difficult to achieve if it is not what many individual men and even many women actually want. Thus, to the problem of overcoming the many religion-based objections to women's equality and the associated natural resistance of any established power group to relinquishing power, we may have to deal with the subtleties of interpersonal relations, including sexual relations, and their effect on personal attitudes toward equal rights for women.

WOMEN'S RIGHTS FOR A BETTER WORLD

Having noted some of the issues that could influence progress toward equal rights for women, there is an extremely powerful argument that favors achieving them. The conclusion of this argument is that reaching that goal will contribute significantly toward solving many of the world's most pressing current problems, leading to what many of us would call a higher stage of civilization. We first review what many of these world problems are, noting that one of them, overpopulation, is exacerbated by the antiquated population policies of tradition-minded religions.

Most thinking people would agree that the risk of large-scale warfare, environmental and ecological degradation, socioeconomic injustice, and overpopulation are major world problems that are not yet solved. The destructive consequences of large-scale warfare are obvious in a world armed with weapons of mass destruction. This subject is taken up again in chapter 8 and is not considered further here.

Much environmental degradation can be regarded as the direct effect of human activities on the environment. Most of this is harmful to human life and many other sentient beings. Climate change driven by global warming is a prime example and is considered here in several chapters. Human activity has been implicated as the major cause by an overwhelming majority of the scientific community. Ecological degradation can be defined as a broader decline of health in the entire ecosystem. The two terms overlap, but the usage of them here differs somewhat. *Ecological degradation* is broader in scope than *environmental degradation*.

Socioeconomic injustice is treated elsewhere in this book, but over-population needs further emphasis, in part because traditional population policies create serious obstacles for realizing equal rights for women. Arguments that planet earth may already be overpopulated and is almost certainly headed in that direction today have been made elsewhere, so it is a structural, or institutional, problem that we examine here.

Using the Roman Catholic Church as an example, but acknowledging that this institutional problem is also serious in other religions, most people are familiar with the resistance of that church to measures that would reduce population growth. Here I ask why this might be true. Instead of asking what is good for the secular lives of people, we can ask why this policy might benefit the Roman Catholic Church. That is the approach I take here.

History demonstrates that the longer an institution exists, the more it develops an institutional interest in all policies that favor its survival and growth. As long as the assumptions these policies are based on are sound, this position has merit, but when conditions have changed, policies often need to be changed as well. There is no reason to assume that the Catholic Church, *viewed as a human institution*, is fundamentally different from any other large institution in this self-perpetuating acquisitive tendency. Other long-established "churches," nations, large empires, and large contemporary multinational corporations behave in many similar ways. It is tempting to expand David Hume's well-known aphorism to say, "Institutionalization is the slave of collective passions."

Given abundant evidence of the problems human overpopulation has caused and is sure to cause in the near future, why would this venerable institution resist population stabilization so stoutly? Traditional Catholic teaching is the obvious short answer, but there may be more. Like most traditional religions, including Protestant fundamentalism, Roman Catholicism is highly patriarchal. To the extent that equality for women will undermine patriarchy, opposition to this equality makes sense for an institution dedicated to preserving it. However, the possibility that an institution devoted to relieving human suffering may also be an inadvertent beneficiary of the very suffering that institution is officially devoted to reducing must also be considered. While it is troubling to think that such a policy is deliberately generated for the purpose of promoting institutional power, we know that institutions can rationalize as much as the people comprising them; and there is no doubt that this church has historically thrived when human suffering has peaked. I doubt there is a *conscious* desire among the leaders of this church or others to promote suffering through their current population policies, but that is a likely consequence.

American Protestant fundamentalist churches have at times also been institutional beneficiaries of human suffering, and this may once again be true today. The population in the contemporary Islamic world is increasing rapidly, and many Muslim countries are not well endowed with agricultural land that could support a large population. The same institutions that provide much of the opposition to women's rights are also the ones promoting population-growth policies, usually justified by a decree claimed to have been received from a patriarchal deity. These institutions continue to exercise great power over people's minds.

The relevance of this to the challenge of gaining full rights for women should be obvious. There are clearly powerful institutions that are opposing it and can be expected to continue opposing it for some time, and these are the very institutions most resistant to change. Their appeal is often based on claims to some degree of infallibility, a claim that is particularly comforting to ignorant people under stress, namely, those who are probably suffering the most.

Thus the opposition to achieving equal rights for women is formidable, comes from many sources—economic as well as religious, and may include subtle psychological factors not yet well understood. In light of this opposition, how strong a case can be made for it? Many have made this case based on the fact that women and men are equally members of the human race, and that simple justice demands equality of rights, even while acknowledging the obvious gender differences. While this position is sound on the basis of equal justice for all people, it has so far proved inadequate to achieve the goal in practice. History suggests that additional arguments will be needed.

The arguments that follow here will emphasize the advantages likely to accrue to humanity overall if we choose to live in a future civilization that honors in practice as well as in principle full rights, respect, and opportunities for all people, women as well as men. Passing fashions aside, most people do not like change for its own sake, and there is much to recommend this attitude. Only when a majority of people in all major cultures think a civilization based on equal human rights will serve their needs better than the one that has developed to date is the change likely to occur. Demonstrating that to people's satisfaction is necessary.

So we ask if equality for women can reduce the threat of major war. I think the answer is unambiguously yes. It was said above how difficult it would be to induce millions of women in two contentious nations to submit as easily as young men to military adventures in which millions are killed and millions of others are maimed. *Millions* may seem an exaggeration, but the twentieth century demonstrates how quickly this number can be achieved. Joseph Stalin is quoted to have said, "A single death is a tragedy. A million deaths is a statistic." One million deaths is 1/28 of the losses suffered by the former USSR during 1941–1945, according to the country's own official records. That many of these deaths were due to starvation does not reduce the number.[21]

Will equality for women reduce environmental and ecological degradation? Here, too, I think the answer is yes. There is abundant evidence that women are more interested in living in harmony with the environment than in exploiting it. Some exploitation is necessary for human betterment, but exploitation becomes destructive when carried to extremes without proper regulation, for which we have many contemporary examples. No sensible person would oppose essential mining or logging operations that provide the minerals and other resources that modern life depends on. However, this does not require destroying mountaintops, creating huge sources of toxic wastes, or clear-cutting large expanses of forests to save a few dollars.

Regarding the ecosystem and the many lives it serves, a number of people have noticed that what is sometimes called "the maternal instinct" seems to exhibit itself in other ways in women who have no children, both through activities that benefit those in need and through a love of pets, animals, and nature in general. Some men share these strong interests too, but I question if the number is as great. I am not aware of a properly conducted statistical study to confirm this observation, but I know many observant people who would agree with it.

Will equality for women reduce socioeconomic injustice? Again, I think the answer is yes. Few would deny that men tend to be more overtly competitive than women, including ferocious competition for money and power, with no regard for others in the most extreme cases. However, before reaching a firm conclusion, the competition among women for mates and social status must also be considered. In an equal-rights society, women may become more competitive and less regarding of others, while the reverse may be true for men. Compassion might track dependency or increased interdependency, and independence could enhance competition.

Nevertheless, it is hard to think that equal rights will not at least promote better interpersonal communication and understanding, both of which are sure to lead to greater concern for equality based on empathy, a common humanity, and the concern for socioeconomic justice that follows.

Will equality for women reduce the current world trend toward overpopulation? Here the answer is *emphatically yes.* The evidence is overwhelming that women who are not pressured to have many children elect to have fewer, especially when they enjoy some degree of economic independence. These fortunate women are in a much better position to fulfill their wishes as mothers for better lives not only for their (fewer) children but for better lives for themselves as well. If "life, liberty, and the pursuit of happiness" means anything, it certainly applies here. The result for society as a whole is a larger number of productive, educated citizens and a much smaller percentage of children born into poverty and neglect. With a world population predicted to exceed ten billion by 2050 CE, surely we can move beyond a population policy adopted by traditional religion, when the total human population was well under 10 percent of this number, to one that emphasizes the quality of human life over additional numbers.

As for the question of resistance to equality for women because of possible male sexual anxieties, we already have a partial answer, even if currently unresolved speculations that support traditional arrangements should prove valid. There is abundant evidence that better communication between people has resolved many such personal issues where they exist. It is not hard to predict that under future conditions of equality and mutual trust, the sexual life for most people will be much improved, and that many people will be thinking about it much less.

On the basis of these comments, and in agreement with most progressives everywhere, I think that the world will be a more civilized home for all our descendants when women are accorded equal rights, opportunities, and respect. Yet, for the reasons discussed in this section, I would not be surprised if sexism proves to be the last major ism to fall before the advance of human rights and human knowledge. Passion only yields to evidence when the evidence is overwhelming and the old ways prove destructive. But there are other passions and other opportunities waiting.

PART 3

TOWARD A MORE

HUMANISTIC WORLD

CHAPTER 7

RELIGION EVOLVING

THE ENLIGHTENMENT AND RELIGION

No thesis on the Enlightenment that fails to consider the challenge it poses to traditional religion could possibly be complete. Historically, this challenge has been thoroughly studied by scholars and is well documented. From Voltaire's famous "*écrasez l'enfâme*" to the passionate attacks on religion by atheist writer Christopher Hitchens, several generations of Western nontheists have attacked religion on grounds ranging from lack of evidence for religious beliefs to outrage over crimes committed in religion's name. Even many nonbelievers who have advocated a modus vivendi with tolerant religion have also predicted its eventual demise as science continues to advance. Thomas Jefferson famously predicted that young men born during his adult years would probably become Unitarians, a tolerant sect that is viewed with disdain by many religious fundamentalists today.

We now know that these predictions, which have been made in every generation since the eighteenth century, have not come to pass. Religion of some form continues to appeal to a large fraction of the human race, if not in most of Europe and East Asia, then certainly in much of the rest of the world, and—surprising to some intellectuals who find America otherwise advanced—also in the United States.

Since religion has assumed many forms throughout recorded history, and, by inference, throughout much of human prehistory, I need to define

127

how this term is used here. In this book, I will define *religion* as a belief in a supernatural realm that has a substantive existence independent of what the human imagination alone conceives. Religion so defined contrasts to the view of the nontheist that, however real this realm may seem to the imagination, it has no substantive reality outside of the human mind. I leave it to the common sense of the reader to grasp what I mean by the words *substantive* and *real,* or else we would need a doctoral dissertation on the meaning of the terms. Suffice to say, to the nontheist, all experiences of gods, demons, or other thrilling and terrifying spiritual manifestations are entirely a product of the human mind. This definition of religion can be expanded in many sophisticated ways, but it will serve our purpose here.

Thus, in agreement with common practice, I consign all gods, demons, ghosts, immortal souls, and other postulated inhabitants of the spirit world to this supernatural realm. What has been said so far in this chapter neither proves nor disproves religious claims. While among the nontheists myself, I will maintain a neutral stance here, though chapter 4 gives a hint to the reasons for my own personal atheism.

Many reasons are offered to explain why supernatural religion still holds a strong appeal for many people. These reasons range from human mortality and the personal hardships many people continue to endure to psychological speculations based on experiments conducted under laboratory conditions. These experiments tease out the possible origin of irrational beliefs for which there is no substantive evidence, based on recent knowledge of how human neural networks develop and function. Some of these experiments were discussed in chapter 5. Nevertheless, I simply note here that religion has so far defied recurring predictions for its impending demise.

Because predictions of religion's imminent demise continue to be made in spite of their failure to date, a considered skepticism over the rapid transformation to a wholly secular society may arise, even among nontheists. This skepticism does not deny the possibility of dramatic changes under certain conditions that could arise and may already have arisen in parts of the world. A popular contemporary example of such dramatic change is Europe. Many surveys show that Europeans are much less religious today than they were before the two world wars of the last century. Further studies indicate that a major reason for this shift is the humane social networks set up in Europe since these wars, which gave people greater personal security and thus reduced their perceived need for religion. While this is likely to be true, it is also reasonable to suggest that the terrible suffering Europeans experienced during and after those wars convinced many of them that their traditional God had failed. Such a reaction would resonate with the recorded response of earlier peoples who destroyed their idols following tribal disasters. Whatever the case, professional historians have long noted that confidently assigning causes to recent events is risky, and that only time will tell if the current European trend away from religion continues.

Few humanists would wish for a global catastrophe that *might* stimulate a sudden transition to a completely secular world, though familiarity with history suggests that major disasters are often the causes of dynamic social change, often moving in directions difficult to predict. Scholarly studies of the Black Death in Europe suggest that the catastrophic reduction of the population due to the plague was a major cause of the Renaissance and the end of the generally bleak Middle Ages. Studies have shown that many of the survivors of the plague resumed the same kind of lives they had lived before it struck, even in villages that lost half their population. However, the impact on theologians and the clergy was apparently more unsettling.[1] Compared to theology in the Middle Ages, the Renaissance that followed was akin to spring following winter. Even in the normally conservative Roman Catholic Church, there was a growing recognition during the Renaissance of the need to experiment toward improving the secular condition of humankind.

I will argue in what follows that one of the most compelling reasons that religion has survived is that religion itself has evolved through time and continues to do so today. Fundamentalism may urge a literal reading of ancient texts and in some cases promote anachronistic doctrines with violence. However, the more subtle theologians of all major world religions have themselves evolved beyond that point, even where their respective faiths give official claims to primacy and may publicly state they are immune to change.[2]

FROM THE BRUTAL GODS TO THE DISAPPEARING GODS

To consider how religion has evolved, it is necessary to recall how rapidly our cultures have changed relative to our genetically based human nature. Compared with the rate of cultural change, and especially with the current rate of scientific and technological change, biological evolution of the human species has been very slow. Thus the differences in what people believed several thousand years ago—near the end of the Paleolithic era—and what we believe today are almost entirely due to changes in our cultures, including our modern scientific culture. This leads directly to the question of how religion has evolved from its earlier stages to its more sophisticated stages, especially in the minds of the more thoughtful. We can start by asking about supernatural belief systems in early hunter-gatherer tribes compared to those in the larger, more complex Axial Age societies Karen Armstrong describes in *A History of God*, or compared to our scientifically advanced societies of today.[3] When allowance is made for the limited knowledge available in earlier eras, were our Paleolithic ancestors any less rational than we are today?

Certainly religion was present at the transition from the late Paleolithic to the Neolithic, a development that occurred when the earth emerged from the last great ice age. This transition took place around ten thousand years ago and gave us the current relatively warm period during which civilization developed in stages to the state it has reached today. Archaeologists and anthropologists continue to conduct studies of how this occurred, and the brief discussion here will skirt many of the important issues they have raised. However, there is almost universal agreement that widespread domestication of animals; agriculture; complex kinship systems; the development of settled communities with divisions of labor; written languages; increasingly complex technologies; and eventually at a very recent stage, modern science all occurred during this period. Here we ask how religion coevolved with these other developments.[4] The answer not only conforms to common sense; it is also confirmed in many studies by scholars of religious history.

Many early religions invoked extremely brutal gods. Such gods can be found in the religions of the Incas, the Mayans, and the Aztecs of the New World, and in the old tribal god of the Israeli people. Human sacrifice was quite popular with these New World religions. It was risky to be an attractive virgin during a Mayan crop failure. The original Jewish Yahweh would be considered a monster by many contemporary theologians, including most modern rabbis. Many early Hindu deities, of whom the number is impressively large, were not altogether pleasant either, especially in their more destructive manifestations. Lord Siva was both the Creator of Worlds as well as their Destroyer, while his consort Kali was the terror of men, as one can see from the hanged male corpses she often wore as earrings. I will refer to these old gods collectively as *the brutal gods*. Important details have been overlooked in this broad generalization, but there is little doubt that many historical old gods exhibited nasty attributes that all but violent fundamentalists among the contemporary religious would deny today.

From cave art, we have evidence that what might be called religion had developed by the late Paleolithic period.[5] It is useful to imagine what the state of mind of an early shaman might have been. These are educated guesses, given the scarcity of reliable information on this early epoch. Yet some writers have speculated that these early gods may have emerged in the course of a shaman's dreams or even during drug-induced hallucinations.[6] Modern science has reduced these experiences to fantasies, but they would have made perfectly good sense to an intelligent Paleolithic shaman. That early gods were often brutal is also reasonable, assuming a growing perception among the more intelligent of these early thinkers that only a punitive deity could likely control the behavior of people living under conditions conducive to starvation, rape, intertribal warfare, and the stresses produced by an unpredictable and occasionally violent natural environment. That these same gods would provide rewards to the faithful in a netherworld after death would give some balance to this early theology, for there must have also been times when even this primitive human life was agreeable. Nevertheless, given the dependency of early tribal humanity on a natural order that could be provident but also highly destructive for reasons people did not grasp, it was not unreasonable for them to assume that if life is often harsh, brutal, and demanding, so must be the gods.

From this perspective, *the compassionate gods* would eventually emerge as conditions of life improved. Jesus and Buddha immediately come to mind, though a true Buddhist might object to the "god" label. It is not hard to imagine why such gods would eventually emerge when life for many people became more secure under the development of fairly stable, settled communities. Yet even then, the need for social coherence and control would likely encourage concepts of deity that retained certain punitive features, while developing further the more humanistic attributes of religion. No doubt this development followed different paths under the conditions people actually experienced in different cultures during the Axial Age. However, it is probably not unreasonable to suggest that as the conditions of daily life became more agreeable, so did the gods.

These improvements in the conditions of life did not eliminate the need for social control, especially of clearly unacceptable behavior. Determining the punishment the new gods would recommend for unacceptable behavior was clearly a major task for the different theologians and legal figures of this era. This no doubt reflected their estimation of how much punishment was needed to promote social harmony. Given the general ignorance and illiteracy of most people in the world at that time, it was probably not difficult to convince the majority of the divine origin of these ecclesiastical pronouncements. Some of these theologians may have actually believed that God was channeling these proscriptions through them. The priestly classes had enormous authority during the Axial Age, and it was often dangerous to dispute their interpretation of "the higher law." The role of religious authorities in contemporary efforts to regulate society is pursued further in the next chapter on politics.

Finally, we come to our own era. If the broad generalizations regarding brutal gods and compassionate gods suggest an evolution toward gods that conform to the general experiences of people living in each epoch, it is tempting to call those studied by the well-educated subtle theologians of the current age *the disappearing gods*. Naturally, there remain many religious people, and some theologians, who would take exception to this characterization. Here I refer only to those modern theologians who are making heroic attempts to reconcile religion with modern science, especially evolutionary biology and cosmology. The arguments of these theologians undoubtedly have value for educated modern people who have abandoned traditional dogmatic beliefs but who still desire the comfort and community aspects of religion. The nontheist who looks upon this with distain may show some lack of humanistic understanding, given that the transformation from a theistic to a nontheistic worldview is not accomplished overnight. Indeed, this is arguably a psychological impossibility. We noted in chapter 5 that changes in our mental concepts have physical analogs in our brains, which do not change instantaneously. Thus as long as a significant fraction of humanity continues to take religion seriously, it hardly seems ethical to oppose tolerant religion for people who are arguably dependent on it for their personal well-being. However, there is another view on that question that needs consideration.

SHOULD NONTHEISTS ATTACK RELIGION?

There is another, more critical view on how secular people should view contemporary modern religion, even in the presence of the rather unobtrusive disappearing gods. This viewpoint takes a more jaundiced view of all religion. It argues that even tolerant religion is harmful because it legitimizes the idea that the supernatural realm exits, independent of the natural order, when there is no reliable (i.e., scientific) evidence to support this position. A number of popular books advocating this position have been written recently.[7] Some of these authors also argue that one particularly harmful consequence of seeking an accommodation with even tolerant religion is to indirectly give aid and comfort to religious extremists. They also offer a critique of the many ways in which even tolerant religion can create confusion and inhibit critical thinking.

I agree with some of these arguments. Nevertheless, it is hard to believe that the tolerant modern theists, who are often very supportive of contemporary science and progressive social policies, are giving much aid and comfort to the advocates of reactionary views and violent behaviors, as the writings of some—though not all—atheist authors suggest. The kind of modern theologian just described in the previous section is usually an opponent of aggression and a champion of cooperation and societal progress.

It would be hard to find a single religious humanist who did not agree with atheist authors *when* their critique is applied to violence-prone religious reactionaries. The suspicion of science and the violent inclinations of many religious fundamentalists contrast sharply with the humanistic views of modern religious progressives. Failure to recognize this and to make a clear distinction here is, in my opinion, a serious mistake. Critics of all religion may also overlook the fact that the more offensive passages in traditional religious texts are usually discounted by most modern theologians and educated religious laypeople today. Much of the brutality recorded in the Old Testament is viewed as the ancient history of a people who were acting by the standards of a more barbaric age. Another dubious tactic of some religious critics is blaming religion for every crime committed in religion's name, when even a superficial knowledge of history demonstrates how often religion is used by secular powers to achieve secular ends. The Christian Crusades of the late Middle Ages offer a good example, even though the stated goal was entirely religious. Nor is our current era exempt from using religion to advance clearly secular agendas.

It is true that earlier religious teachings that postulate angry, vindictive gods are still found among the violent fringes of Islam in the Middle East and certain Protestant fundamentalist sects in America today, as well as fringe groups in Judaism. Other faiths include a (usually small) fraction of fanatics as well. However, the full contemporary religious picture is demonstrably more complicated than that presented by those who would eliminate religion entirely today if they could. Speaking as a secular humanist, I note that secular thinkers have still failed to show how a completely secular society can satisfy those basic human needs that continue to attract many people to religion today. This is a formidable challenge, and the existence of ideas popular in the academic world that have yet to work in practice only demonstrates the point.

Philosopher of science Daniel C. Dennett, though sometimes identified with "the new atheism," actually exhibits a highly nuanced view of religion that is not entirely unfriendly to it, on the grounds that religion may
still be a necessary part of what holds many societies together at the current
stage of human societal evolution.[8] Thus Dennett's believing in (the value
of) belief resonates with the attitude of an enlightened minority of deists
during the Enlightenment. Such a view may accurately reflect existential
realities of that era and of ours today, as well.

Many contemporary theologians outside of fundamentalist circles
have developed highly abstract concepts of deity, which they identify
with esoteric concepts like "the apparent perfection of mathematics" or
"the quantum fluctuation in a vacuum that brings a universe into being."
(Don't try to imagine that.) I once heard such arguments from a Harvard
Unitarian theologian speaking at the New York Academy of Sciences. These
theologians are clearly trying to be reasonable, even if their concepts of
deity are not what many parishioners are expecting, or what the same theologians might tell a large congregation of the faithful. That even the above
statements can be questioned by well-trained mathematicians or require
interpretation by competent cosmologists does not mean that modern theology is wrong in its effort to provide a compassionate service to a world
where widespread ignorance dominates sophisticated intelligence, leaving
a large population that still yearns for existential answers.

The case can be made that we are witnessing a historical process that
is far from played out, even though the modern God seems, like the unfortunate climber of Everest who vanished into the mists, to be rapidly disappearing into thin air. As science continues to advance, there are those of us
who think the best strategy is to let this natural evolution go to completion
while always confronting and blocking regressions along the way.

As an example of how difficult it may be to eradicate religious superstition quickly, we note how easily false beliefs can be established in the highly
impressionable minds of young children. Chapter 5 gave one example
based on a scientific experiment. Atheist author Richard Dawkins has com-
mented on this phenomenon. He argues that a way should be found to
counter the generally recognized right of parents to transmit their own

religion to their children.[9] Yet even Dawkins recognizes that this would be difficult to accomplish in practice. Most progressives would oppose such state-sponsored regimentation today. Many people might well forego parenthood altogether if they lost that much control over their own children. Finding a solution to this problem will be difficult, and we return to it in chapter 10, where education is discussed further.

It is easy to demonstrate why the problem of religious indoctrination of the young is serious. Traditional religion has long recognized the vulnerability of the young and has used this knowledge to indoctrinate children before they reach the age of reason. Most secular humanists would argue on ethical grounds against this practice of many churches today, seeing it as a way of denying the young person the right to form her own views on religion only after she has reached the age of reason. Yet it continues to be a common practice. Our brains develop most rapidly when we are young, and the impressions formed then tend to be the most enduring. Once indoctrinated at an early age, restructuring the neural networks that are the basis of these impressions becomes increasingly difficult. This probably explains why the most comfortable atheists are often those who have never received religious training. In contrast, those who received religious training when young, and later rebelled against it, often seem the angriest of all, and the ones most interested in suppressing religion.

Before concluding that this practice of early indoctrination has always been unethical, it is necessary to consider the conditions of life during earlier historical and prehistorical epochs. We need to consider again how our progenitors with physical brains very like our own reacted to the challenges of those earlier times, when life for most people was more difficult than it is for many of us today. Indoctrinating young members of a tribe in its religion at the stage in life when superstition was most likely to take hold probably gave that tribe and its members distinct survival advantages, especially at a time when small-group survival was critical and precarious. If we then add to this the inertial forces of tradition and habit that operate in all societies, the persistence of this parochial tendency is not difficult to grasp.

To overcome any kind of indoctrination, we are left with the need for widespread public education that includes science and critical thinking.

Unfortunately, at this moment in the United States, the trend may be moving in the opposite direction. Homeschooling has been advocated by numerous religious fundamentalists as the best way to promote nonscientific views such as creationism and to protect young fundamentalists from what is erroneously called "the religion of secular humanism." These children will almost surely get an inferior education.

Despite these developments to promote indoctrination and restrict early education, history confirms that, outside of reactionary circles, concepts of God and religion have changed and have become generally more refined as knowledge grows. Even the relatively conservative Catholic Church in the West is continuously modifying its interpretations of natural phenomena to bring dogma into closer conformity with reliable scientific knowledge. There are limits to this endeavor as long as one is committed to certain dogmas, but that major changes have occurred during even this venerable institution's history has been well documented. There is no reason to think that this long-established evolution of religion will stop at any time in the future.

However, whether a person living today would recognize the future form religion might take is another question. Even the term itself could eventually be redefined. One major insight biological evolution has given us is that human beings are adaptable and need to be in order to survive. An evolved form of religion that is based entirely on reliable knowledge that is subject to empirical confirmation may not be out of the question.

As for the question that began this section—should nontheists attack religion—in the general case I would argue that we should not. It does not serve the ethical outlook of the Enlightenment to attack the nonviolent viewpoints of tolerant religion with an effort to suppress it. (Intolerant and violence-prone religion is another matter.) Many of us who stand with science and reason are content to pursue that approach, confident that it will prove the best way to resolve most of the fundamental issues in time, and equally confident that the process is unlikely to proceed quickly. That does not mean that religious claims should not be subjected to the demand that "extraordinary claims require extraordinary proof." We certainly think that they should be. In addition, religion has no right to insist on special privilege in the public square, and any demands for that should be resisted.

THE RELIGIOUS CULTURE WAR
IN AMERICA TODAY

One development already noted is the current conflict between reactionary dogmatic religion and those who actively oppose it. This conflict is occurring with varying intensity over large parts of the contemporary world. Many of us see this as a crisis for and within religion but do not predict a precipitous decline for the institution because of it. History shows repeatedly that when a new social paradigm is introduced and becomes well-known, a strong conservative reaction inevitably occurs. It is obvious this is occurring in America today, as millions of religious people have awakened to the growing secularization of society, and many of them are threatened by it. Ideally, when such opposing positions emerge in a democratic society, a peaceful synthesis that incorporates the best elements of both positions eventually emerges. Unfortunately, such a reasonable outcome is not guaranteed but depends on the goodwill of all major parties in the dispute. This does not mean the final outcome should always average the differences between the opposing positions. One side may receive the greater weight in the resolution.

In considering a practicable solution, a large majority of generally tolerant Americans would probably support institutional separation of church and state as a reasonable compromise. Such a compromise does not exclude citizens from advocating legislative positions on issues that reflect their personal philosophy and/or religion. Indeed, this right is guaranteed, short of violating the establishment clause of the First Amendment to the American Constitution. However, even this First Amendment principle has never been fully realized, and those who wish to compromise it further are once again very active today. It may be true that religious fundamentalists are more insecure when confronting complex issues than are the tolerant religious and the nontheists, but this may be irrelevant to the unwillingness of religious fundamentalists to accept a reasonable civil compromise, especially when high ecclesiastical and legal authorities encourage clearly parochial religious civil programs. The issue continues to generate considerable "heat."

The primary antagonists in the contemporary religious culture war are those who insist on literal interpretations of ancient religious texts on one hand, and an increasingly concerned population of scientifically literate nonbelieving and other open-minded people on the other. The latter group in America had previously adopted a live-and-let-live approach through much of the twentieth century, fortified by a reasonable degree of institutional church-state separation. These nontheists and many religious liberals have since become disturbed by the increasingly aggressive approach of religious fundamentalists to break down this separation and move American society in a more theocratic direction. There is a tendency for each side, like children in a schoolyard fight, to assert that the other side "started it." Notwithstanding the false claim that secular humanism is a religion, the fundamentalists assert that liberals have tried to impose "the religion of secular humanism" on America. The nontheists and religious liberals have responded that the struggle began in earnest only when transparent attempts to break down the wall of separation between church and state started to score some executive and judicial successes within the federal government.

Although the war terminology is unfortunate, by suggesting an intense dislike of "the other" that can eventually inspire more than verbal attacks, the disagreement has so far been moderately dignified and nonviolent, at least in the United States. Nevertheless, it may not be an exaggeration to say that battle lines seem to have been drawn between uncompromising religious extremists and those who detest the idea of any kind of theocratic republic. The former side is demonstrably well funded in spite of the relative poverty of many of its most ardent supporters. Corporate America has struck some political deals with those who favor limited theocracy, as will be noted in the next chapter on politics, but it has done so nervously and with some legitimate concern for the future. The latter side enjoys formidable intellectual firepower in academic and scientific circles, but until recently it has been less vocal in the greater society.

Twice before in American history there have been major attempts to transform the country into a citizenry of passionate believers who would build the "Shining City on a Hill" and lead the world to universal Christian (or perhaps Judeo-Christian) victory. Twice before this has failed, in part from fizzling out as great issues of more immediate concern arose.[10] This could well happen again, but as of this writing the religious culture war described here is real and does not seem likely to dissipate quickly on either side. Both sides like to point to their favorite heroic figure, whether Jesus or Jefferson, often overlooking the latter deist's favorable written comments on many of the former's principles of personal morality. As long as this situation exists, tolerance is more difficult to achieve in the public square. This is clearly not a healthy situation for a democratic society that depends on mutual respect and a willingness to compromise.

Fortunately, there is a third group of people in the United States who may slightly favor one side or the other but who normally prefer not to be inconvenienced by a struggle and do not wish to become personally involved. As usual throughout history in troubled times, this is the largest group of people. Because these people are primarily concerned with their personal lives, many of them feel vulnerable when faced with major societal developments and often avoid taking sides until they can assess who seems to be winning. Normally, they can be counted on to be the least active group. Many of these people are skeptical of traditional religious claims, but they may also be reluctant to cut their ties with religion for a variety of personal, familial, or even professional/vocational reasons. Most of them want a peaceful, productive existence and will avoid divisive issues that fail to affect them directly. A significant fraction of these people in America today are relatively nondogmatic in their religious beliefs and tolerant in their political outlook; but they remain quiet about it.

This large group in the middle is the key to which way the entire society will go in the future, as it has been in the past. Whatever its faults, America remains a functioning democratic society as this religious culture war continues today. For this reason, both religious fundamentalists and secular thinkers are courting this majority that lies in the middle. A recent book on the United States Supreme Court, *The Nine*, argues that the differences between the relatively liberal Warren Court and the relatively conservative current Court reflect shifting moods in American popular opinion more than arbitrary decisions by activist judges.[11] This recent increase in Supreme

Court judicial conservatism corresponds to the documented increase in political influence of religious fundamentalists in America today and to a weakening of the wall of separation between church and state. The crucial question is how this still-largest group in the middle will react to these developments. The answer could come soon.

With these final comments on the evolution of religion in America today, the next chapter moves on to politics. While it may be impossible to completely separate politics from religion in practice, that was one of the goals of the Enlightenment. It was the ideas of the Enlightenment and their knowledge of the destructive religious wars of the Reformation that inspired Jefferson, Madison, and others to favor the complete institutional separation of church and state. This is reflected in their writings and ultimately led to the famous establishment clause of the First Amendment to the American Constitution, which also provided for the free exercise of religion that has been partially responsible for the success of numerous religious dominations in America today. That full institutional separation of church and state has proved impossible to achieve in practice undoubtedly reflects that many Americans have resisted the idea from the start. The idea was clearly too radical for many people to accept when the Bill of Rights was written, and it remains so today.

It is in the realm of politics where this issue will probably be settled in America. If one major goal of the Enlightenment is to be reached, achieving human betterment for all peoples everywhere independent of the parochialism of particular religions, it is in the political realm where this will probably take place.

CHAPTER 8

THE PRIMACY OF POLITICS TODAY

THE ROLE OF RELIGION IN POLITICS

*P*olitics and religion have always been mixed, and this situation remains true today. One obvious reason for this is that is what many people still want, as demonstrated in America by the reluctance of millions of citizens to favor complete institutional separation of church and state. Whatever the First Amendment may say in the famous establishment clause, many Americans feel insecure without constant reference to God in the political realm.

Politics and religion have been interwoven from the beginning of recorded history. There is abundant anthropological evidence that the two developed together well before the written word. The previous chapter noted the role of shamans in early cultures. While these individuals may not have been the political leaders of their tribal communities, they played a large role in influencing major decisions in times of crisis. Those who could claim contact with a spirit world, then almost universally accepted, were sure to have had enormous influence on tribal decisions.

Moving forward in time, we note many examples of this spiritual influence throughout Western history, especially during the Middle Ages, a period of political fragmentation and impoverished serfdom. The Roman Catholic Church was the single overarching authority during this period.

143

Even the powerful Holy Roman Emperor Henry IV saw fit to submit to papal authority at Canossa in 1077 CE, though the event gained important secular advantages for the penitent, who eventually succeeded in removing the same pope. Church and state vied for primacy in the medieval world, but even the most powerful secular authority maneuvered to establish a friendly pope in Rome, since papal influence transcended political boundaries.

Medieval history suggests that when secular conditions of life are harsh, people may turn to religion, even authoritarian religion, if it offers personal hope and some semblance of social order. Ignorance, poverty, brutality, and a conviction that humanity had degenerated from its classical heights were ubiquitous during this period of Western history.[1] When the likely reaction of a people to those conditions is considered, it is not hard to grasp the appeal of the medieval Catholic Church to many of them during that era. Nor has this state of affairs been overcome everywhere today. It is impossible not to see some relevance to parts of the contemporary Islamic world, though the authority structure of the two religions differs widely.

Authoritarianism in religion creates a mind-set hardly conducive to democracy in politics. The struggle toward political democracy demands that certain minimum conditions exist in any coherent society. Should these conditions fail to emerge or cease to exist, democracy becomes far more difficult to establish or maintain. We are learning, arguably the hard way, that American-style democracy may not work without significant modifications in certain Islamic countries today. (In fairness, the British seem to have learned similar lessons earlier, also the hard way.) The detailed conditions for democracy undoubtedly vary according to local cultural circumstances. There is evidence that the New England town meeting was inspired in part by contact with not always friendly, but often democratic, American Indian tribes.[2]

Granting many cultural variations, it is still hard to imagine a democratic theocracy, since the ultimate authority in a theocracy rests with a minority who presume to know a divine law that arose independent of empirical evidence. Or so it is claimed in a theocracy. The writers of the American Constitution were deeply influenced by the Enlightenment goal of removing political authority from any priesthood. Contrary to claims of certain contemporary American writers of the religious and political Right, the founders, while often non-dogmatically religious, were determined to minimize the role of organized religious institutions in the new American government.

Despite this early effort to achieve institutional separation of church and state, reference to God has been reintroduced into American political life many times since, often with popular enthusiasm. Examples include inscribing "In God We Trust" onto American currency and incorporating "under God" into the Pledge of Allegiance. In both cases, these measures were taken during periods of national insecurity. Nevertheless, the Pledge of Allegiance was recited in most American public schools during World War II, when it did not contain the phrase "under God." It was not necessary for Allied victory in that conflict, while soldiers of the defeated German Army were assured of divine favor by the motto on their belt buckles, *Gott mit Uns* ("God with us"). Another example closer to home for Americans concerns the former short-lived Confederate States of America.[3] Quickly adopting a constitution in Montgomery, Alabama, the Confederacy simply copied the existing US Constitution with only four changes. One change incorporated reliance on "Almighty God" into the rebel version. Notwithstanding this supplication to a presumed powerful ally, the Confederate revolt, like Hitler's plan for world domination, failed. Yet when the Red Scare of the 1950s swept across America, the pledge was modified to ensure divine favor for the United States. It is hard to imagine this was a factor in the ultimate collapse of the former Soviet Union. Economic collapse of a cumbersome authoritarian bureaucracy, possibly accelerated by deliberate American policy, is usually argued instead.[4]

These efforts to curry divine favor suggest how insecure the average American must still feel today, even though living in the wealthiest, most bountiful, militarily most powerful country in the modern world. Since we have no reason to think the average American is less able than the average citizen of any other nation, this suggests a high degree of personal insecurity among people in general. If democracy requires a reasonably confident citizenry for its success, it remains vulnerable to people's individual and collective fears, a fact one of the greatest American presidents addressed boldly in 1933 to build people's confidence during a severe global depression. In his first inaugural address, President Franklin Roosevelt noted that the only thing Americans had to fear was fear itself, a sentiment worthy of a free people anywhere. Lack of confidence in a people opens the political field to authoritarian leadership that often associates itself with religious reactionaries. When this happens, progressive political leaders may need to make concessions to the churches if the body politic demands it, even if the legal grounds seem shaky. Despite democratic America's strong global position entering the twenty-first century, the recent movement toward greater church-state entanglement seems to have had broad popular support, suggesting widespread personal insecurity, ignorance, and fear of the modern world, or all these together.

Another American example is provided by certain actions taken to this date (2011) by progressive president Barack Obama. While this president has publically committed himself to upholding the institutional separation of church and state, he has also further strengthened the highly questionable "charitable choice" program begun under President Clinton and expanded under President G. W. Bush. Of course, the legally trained who favor this program have fashioned arguments to justify it, but that is not the point. Constitutions and Bills of Rights are important safeguards of basic rights and principles of operation, but they require interpretation. It was noted in the previous chapter that these documents can be and are bent to accommodate the will of the people in a given era. Alexis de Tocqueville's famous argument cautioning against the tyranny of the majority immediately comes to mind. In politics, what influences public opinion then becomes a major determinant of government decisions, and laws can always be variously interpreted by skilled lawyers or can be and sometimes are simply unenforced.

In one sense, the people rule in every society, so the deeper question becomes what rules the minds of the people. Answering this question makes clear why religion continues to struggle for primacy in the realm of the mind. Those who think this powerful traditional influence will disappear quickly with the rise of science, as many early Enlightenment thinkers predicted, may wish to reconsider how quickly this can be expected to occur. We might ask if it is wise to forcefully advocate what many people are not ready to accept and, if provoked, are likely to react against with hostility. Vigorous promotion of atheism was discussed in chapter 7 and was not encouraged. Some of us would advocate a more subtle, gradualist approach, along with improving education, especially education designed to stimulate critical thinking and knowledge of modern science.

To conclude these remarks on the entanglement of the religious and political establishments in America, we can note the current efforts of the religious and political right to introduce still more roles for religious organizations into government-funded public life today. Charitable choice was noted as one current example. Especially egregious to church-state separationists is the 2002 Supreme Court ruling, *Zelman v. Simmons-Harris*, which permitted taxpayer funds to be used in support of parochial schools, provided the parents—not the state—chose the school. Among the arguments in support of this ruling is that government funding will not be used directly to support religious education. Directly they will not be so used; indirectly they most certainly will. While seemingly reasonable on the surface, this argument is appallingly flawed. The money saved by applying public funding to teach only secular subjects saves the rest for religious education, allowing a reduction in tuition and thus indirectly promoting religion with taxpayer monies. However, to understand even that simple argument requires a bit of elementary critical thinking.

While only the most politically reactionary of religious Americans seem to actually want the country transformed into a theocracy, there are many who wish significantly more theocratic control of American politics. Today, the most vocal group may be the Protestant fundamentalists. Fortunately, American fundamentalists argue so much among themselves that their long-term impact may eventually fade. This has occurred twice before in episodes of eruptive religious enthusiasm in America, noted briefly in the preceding chapter.

If Protestant fundamentalism is the most vocal religious group in America today, there is another religious establishment that has a longer history, and that may have a more impressive record for patience and adaptive politics. Demographics suggest that the Roman Catholic Church could become the dominant religion in America before the end of this century, especially if Protestant fundamentalism eventually declines, as mainline Protestantism already has. To grasp the growing power and influence of the higher orders of this church in America today, one needs only to look at the current Supreme Court of 2011. This Court now has a Catholic majority of justices and may be the most reactionary Court in the nation's history on church-state issues.

One justice in particular exemplifies the power and persistence of this remarkable church. Antonin Scalia is a brilliant lawyer and a devout traditional Roman Catholic. Examination of his decisions, which are a matter of public record, reveals an affinity for traditional views often associated with pronouncements of the higher Catholic hierarchy. Granting the justice's dedication and integrity, many of us would be more comfortable had he been appointed America's papal envoy, assuming that the country must have one. Justice Scalia would have served well in that capacity with the convictions to which he is freely entitled in his personal life. We must still ask if they should be more generally imposed through judicial interpretations.[5]

These things said, it is not the primary purpose here to denigrate any particular religion, only to demonstrate that *religion remains inextricably involved with American government today, not just on the level of personal conviction but on the institutional level as well.* The American Catholic Church had reason to create its own parochial school system in the nineteenth century, as many public schools of that era openly advocated Protestantism in clear violation of the First Amendment's establishment clause. Nor is this an intended criticism of people in general but is rather an observation that humanity in the large, while still arguably struggling toward Enlightenment goals, remains too ignorant and too threatened by crises, real and imagined, to achieve many of these goals today. For that reason, we might expect religion to continue to be with us for the foreseeable future, since that is exactly what large numbers of people have demonstrated that they want, especially when society passes through a difficult period.

Thus, as noted in the previous chapter, the survival of religious superstition in this age of scientific advances and "high tech" can itself be considered an evolutionary phenomenon. The process is still working itself through a population that represents a wide range in personal beliefs and scientific education. It should not be surprising that genuine societal progress might best be viewed as a crawl, always threatened by regression. The broad thesis presented here is that whatever works best will eventually be adopted by people determined to survive and, if possible, to thrive. I am making the case in this book that what ultimately works best is the reliable knowledge of science. I am also asserting that universal adoption of this is unlikely to happen quickly.

The persistence of religious superstition need not continue indefinitely. It is not unreasonable to assume that with the continuing advance of science combined with evolved human aspirations, religion will either change radically in nature from its more traditional forms or it could die entirely. I can offer no guess as to which of these alternatives is more likely.

THE PRIMACY OF POLITICS TODAY

Despite persistent efforts by some religious leaders to urge conformity of government policies to their religious ideologies, the political institution has gained influence at the expense of organized religion over much of the world today. This shift took place when a better secular life became available to more people, and secular Enlightenment goals have slowly gained weight in their priorities. Notwithstanding the challenges traditional religion is making to the political establishment today, the latter has now become the major concern of most educated people.

Wealth is now distributed and redistributed far more by governments than by churches, synagogues, temples, and mosques. For all the challenges democracies face, the idea of government by law, not by powerful men or gods, continues to gain new adherents. Jurisprudence authority John Rawls, also mentioned in chapters 5 and 6, invokes the ethical principle of fairness to all members of a society as the democratic basis for the just legal system that he advocates.[6]

Police and military power also lie almost entirely with the world's governments today, not with religious institutions. Nongovernmental terrorist activities are sometimes cited as a third new global power broker, but such activities are the strategy of a desperate people who lack other means for achieving their goals. When terrorist activities achieve a measure of success, the societies that promote terrorism are likely to suffer more than the victims, since force majeure remains with the more highly developed nations. We already have examples of isolation or active suppression of those nations that tolerate or support terrorists. Regarding influence in the political process, military forces have learned their own ways to lobby, even in a democracy where the process needs to be somewhat indirect, and especially when the standing force is large. Many thinking people have noticed that military budgets are never large enough for military purposes, even when they are huge. This no doubt reflects how seriously conscientious officers take their responsibility to defend the country, but it can also reflect other considerations.[7]

Large multinational corporations are sometimes cited as competitors to governments. Yet they also recognize their dependence on government and vie for political influence, forming alliances and distributing large funds for lobbying to further their ends. Powerful corporations clearly bear watching in their obvious effort to influence government to maximize their own profits. Granting corporations the same rights as individuals in a century-old American Supreme Court ruling would seem a bizarre interpretation of a Bill of Rights crafted to protect individuals. Perhaps more disturbing to political progressives is the 2010 Supreme Court ruling *Citizens United v. Federal Election Commission*. This ruling relieved corporations of many restraints on funding political campaigns and increased their potential political influence. Unions were also exempted, but there is little doubt where the larger budgets for lobbying lie. The effort of wealthy corporations to favor such rulings demonstrates how highly they regard political influence. It is hard to imagine the CEO of ExxonMobil paying as much attention to a proclamation of the pope on global warming as he would to the threat of new government regulations on greenhouse gases.

Thus it can hardly be said that corporate ethics are guided primarily by religion, or that they operate independent of government regulation, much as a few may wish. The very effort of powerful corporations to influence the political process demonstrates their dependence on it.

If we add to the religious, military, and corporate attempts to influence politics, the emergence of the almost-uncontrollable global Internet along with a media that reaches huge audiences, we are confronted with a rich stew of influences. Whether all these often-competing, and sometimes-cooperating, interests will largely cancel out one another in influence or whether a new dominant alliance will emerge is difficult to predict in a rapidly changing world. One thing is clear. The ultimate goal of all power groups quickly becomes gaining a major degree of control of the political apparatus, even if this is done indirectly. Ironically, this suggests progress toward one of the Enlightenment's main goals, to give the political institution primacy . . . as long as it is both democratic in concept and effective in practice. The further caveat that must be added is the need for a spirit of goodwill among the different political factions. It is generally accepted that democracy will work only if the competing interests are willing to accept compromises that suit no party perfectly.

A final feature favoring the enhanced power of the political establishment is the lessening of religion-based psychic terror in many parts of the world today, though unfortunately some religious groups persist in it. Noting the evolution away from cruel gods to more compassionate gods—and, more recently, to disappearing gods, the fraction of educated people who take seriously the idea of cosmic torments is small, even if the earlier view still commands a following that cannot always be ignored. As fewer people are influenced by threats of divine punishment, most of them become increasingly independent of religion in both their personal lives and their political opinions. This process undermines the power of a clergy and advances society more rapidly toward realizing Enlightenment goals.

The decreasing role of religion in politics in most democratic societies has led to a greater emphasis on individual rights, an important difference that separates democratic societies from more traditional authoritarian ones. We saw in chapter 6 that a major emphasis on individual rights may be a relatively recent development in history and can be considered one of the chief features of the Enlightenment outlook. Advances in human rights invariably lead to support for a democratic political system. Given the power of modern political establishments, human rights are best protected when the people choose their leaders, unless the people are too traumatized to think. There have been exceptions.

Finally, in spite of progress, all modern societies still retain features in common with earlier tribal societies. At least one of these features can interfere with realizing universal democratic goals. *Before considering how a more harmonious global community can emerge, the worldwide problem of contemporary nationalism must be addressed.*

THE PROBLEM OF NATIONALISM

Some scholars have argued that biblical Israel affords the best example of early nationalism, while Germany under Hitler may be the most aggressive modern example. Other historians have noted that a strong nationalistic spirit swept over most of the West in the nineteenth century.[8] *Nationalism*, as the term is used here, consists of in-group solidarity, often combined with out-group hostility and strong feelings of national exceptionalism that go beyond features unique to any well-developed national culture. Territorial expansion and/or imperialism, militarism, and political romanticism represented by grand national myths (the chosen people, the master race, the workers' paradise, the shining city on the hill, etc.) are common features of strong nationalism. The combination of features that define nationalism may vary from one nation to another, but some of these elements are usually present.

The well-studied German case is attributed to many factors, including wounded pride and anger over the many historical depredations the Germans suffered at the hands of their formerly better-united neighbors, especially during the Thirty Years' War of the sixteenth century. To be held solely responsible for World War I in the last century and to lose half their population as Europe's battlefield in the earlier conflict was a cruel blow to a people whose achievements in the sciences, technology, and music are legendary. In addition, German thought often rankled from the prestige of Enlightenment views associated with the rationalism of the primarily French Enlightenment. This and other endemic cultural factors produced a passionate and sometimes-primitive romantic reaction that began in the arts with writers like Johann Gottfried Herder and culminated in the dictatorship of Adolf Hitler that was at once revolutionary and profoundly reactionary.[9] Wildly hysterical enthusiasm characterized the political manifestation of this nationalistic movement, offering a sobering example of how destructive *Sturm und Drang* ("Storm and Struggle") can be when romanticism in the arts is transferred into politics.

American politics are not excepted from the evils of nationalism. The G. W. Bush administration promoted a policy of forcefully exporting American-style democracy to nations that were, and may remain, reluctant to accept the hegemony of a dominant military power with additional agendas to pursue. To the extent that policies are perceived to be narrowly nationalistic, even in democratic countries, efforts by these societies to export democracy seem unlikely to succeed. People everywhere may be relatively easy to fool for a time, but transparently self-serving national policies are hard to hide for long. In the absence of effective global government, history suggests that no sovereign nation can do better than to follow a policy of enlightened self-interest. Lacking an enlightened administration, the situation can quickly become worse.

Recent research provides some evidence for the in-group ethnicity that nationalism thrives on. Ironically, the neurochemical oxytocin that promotes empathy between people and that can induce sacrifice favoring those in our "in-group" may also promote aggressive behavior toward members of a threatening "out-group."[10] This propensity may have conferred Darwinian advantages on small tribes during an earlier phase of human development. Its role in the modern, relatively crowded human world of today could reverse that early advantage. This and other genetically induced propensities of human beings need to be better understood. Our current inability to fully understand and manage our genetic attributes still stands in the way of progress toward a more peaceful global society. To some degree, nationalism is modern tribalism.

Cultural myths can also contribute to aggressive nationalism. All national communities, like religions, tell stories about themselves. A particularly dangerous story that invariably involves national origins and unique virtues is the myth of national exceptionalism. Exceptionalism goes beyond uniqueness, a quality all coherent communities can justly claim. This myth is particularly evident in powerful nations enjoying periods of success, though it may persist for centuries thereafter, like the ultraorthodox Jewish myth that God gave all of Israel to the Jewish people for all time. The Axis nations of the twentieth century all developed myths of their intrinsic superiority, of which Nazi racism was particularly ugly. Exceptionalism is also commonly found in small tribal societies. The advantages to any coherent society are obvious, and until recently they may have been a necessary condition for a community's survival.

Analyzing the many sources of national exceptionalism lies beyond this treatment, but we note its transparently harmful effect on any attempt to achieve a rational world political order. To the extent that exceptionalism (tribalism, ethnicity, etc.) is a natural tendency of any coherent human community, achieving a just world order will not succeed until the issue is confronted and better ways to overcome it are developed. This is one goal of the Enlightenment that remains elusive today, though thoughtful people continue to work toward it. Once again, a better scientific understanding of people, and what triggers our individual and collective behaviors, would contribute greatly to solving these problems.

Extreme nationalism is often associated with aggressive militarism. Particularly instructive examples of the dangers of militarism to the well-being of a society that supports it are Germany and Japan during the late nineteenth and early twentieth centuries. Militarism proved temporarily beneficial to these countries in its early stages but finally brought great suffering to the countries' own peoples in addition to what it inflicted elsewhere. The modern history of both countries shows an antithetical relationship between democracy and militarism. The failure of the democratically inclined Frankfurt Convention in Germany in 1848 ushered in an era of Prussian domination with its militaristic tendencies. Exploited by Hitler after Germany experienced severe losses in World War I, ardent militarism did not shield the outstanding German armed forces from total defeat in World War II, accompanied by terrible suffering there and horrendous crimes committed throughout Europe. Japanese history reveals a strong democratic movement following World War I that was eventually suppressed in the 1930s.[11] Assassination of democratically inclined leaders and rampant militarism in that period led to spectacular early victories against China and the Allies in World War II, followed by overwhelming defeat, almost total destruction of Japanese industry, and huge casualties among the nation's soldiers and civilians.[12] In these two examples, neither democracy and human rights nor the general welfare of the people were served well by extreme popular notions of national exceptionalism and the chauvinistic militarism that accompanied it.

The almost universal tendency of powerful nations to promote national exceptionalism in tandem with a powerful military runs directly counter to the ideals of the Enlightenment, which are global in their reach. Yet, despite examples of the dangers of these chauvinistic views, many political realists seem to favor it. There is a hint of this view in the late Samuel P. Huntington's *The Clash of Civilizations*, which contrasts with the somewhat more visionary notions of "soft power" advocated by his Harvard colleague Joseph Nye.[13]

It is hard not to favor the realist position if history alone is the guide. Self-described realists can make a strong case that human nature, though only partially understood by science today, will impose certain constraints on what humans are capable of. Thus, like it or not, some will argue, ethnicity and competition for resources are in our nature, so we should accept that and do the best we can. Like the alleged Chinese mandarin idea that all history is a cycle of repetitions on a large scale, the realists will cite impressive evidence from history that appears to validate this thesis. This is sometimes translated into constant preparation for war. Huntington's work is highly respected in the foreign-policy community, but an educated nonspecialist reading his work can find it appallingly lacking in an encouraging vision for the future.

We can ask ourselves if there is any way to avoid the kind of depressing and repetitive series of conflicts on increasingly larger scales that are suggested by Huntington. He has made a strong case. Is there any recent major development in history that was not present in earlier eras? Religion and technology both fail that test. Religion has supported both great progressive movements and highly violent regressive ones, and continues to do so today. Technology has given us increasingly useful inventions and personal comforts, and through medicine has yielded great strides in improving our health; but misuse of technology has made the entire world more dangerous for all of us, collectively, than ever before. Even politics has so far failed to produce a peaceful and stable world order, despite the increasing number of democratic governments and at least formal recognition of essential human rights.

I conclude that the only relatively recent development that offers hope for significant further societal progress today is science, with special emphasis on the science of the human being, and particularly more intensive studies of the human mind/brain.[14] Readers might ask if they can think of any other relatively new development that offers equal hope for realizing the global Enlightenment goals already discussed. This conclusion leads to an even stronger conviction that nationalism is an antiquated phenomenon unworthy of educated people living in the twenty-first century. Like religion and technology, nationalism has been well tested, and it has failed. This does not mean the transition to a science-based world will be easy or rapid, but the case for it is strong.

The reasons for nationalism are many and complex, but ego identification with the nation is certainly one. That the more fortunate person can derive strength and confidence from ethically based personal actions and subsequent self-esteem may not occur to one who derives his satisfaction primarily from identification with a group he is told is intrinsically superior. Among other things, this makes the believer more vulnerable to political manipulation.

ACHIEVING A WORKING DEMOCRACY—
WITH AN AMERICAN EXAMPLE

American politics in the first years of the new millennium took many progressives here by surprise. These years remain difficult to assess as we continue to experience the consequences of a reactionary populist movement. Most progressives have been forced to admit we greatly underestimated the influence of reactionary tendencies in this country. The complex of interests this political movement to the Right represents range from dogmatic religion and simple greed to much more practical and legitimate concerns over family stability and jobs. Many of these concerns lend themselves to cynical manipulation of politics, if the people remain ignorant of a larger picture. As of this writing, there appears to be a near standoff between progressives and reactionaries in American politics, with a few hints that real change may be coming.

Some of the conditions for effective democracy are obvious, including a reasonably well-educated, civic-minded electorate and a decent standard of living. We know that Jefferson thought church-state separation was another essential requirement for a successful American democracy, though so far the nation has survived with that principle compromised and constantly under attack from the Right. When Tocqueville wrote *Democracy in America*, he offered some interesting suggestions on religion for the new democracy—as we noted before—suggesting that his own religion, Catholicism, would likely prosper in America, as it has, but recommending that the priesthood stay out of politics and largely out of the personal lives of the people.[15] This seems to have been partly realized, as studies show that a majority of American Catholics have ignored papal exhortations against birth control and seem to prefer public to parochial schools.[16] In contrast, the higher Catholic hierarchy has been generally more traditional and not always friendly to the spirit of institutional church-state separation. From the democratic viewpoint, it seems Tocqueville was largely right about the prospects for his religion as practiced by the American Catholic laity. However, many bishops seem to be waiting for the opportunity to change that.

Individual rights were also a prominent feature of early American political thought. The Bill of Rights was introduced to address this concern, and it is hard to find anything in early American writings that suggest large wealthy corporations need similar protection. However, conservatives' fear of "creeping socialism" was not the concern then as it seems to be again in some circles today. Do we really think the wealthiest nongovernmental institutions in the society need additional protection?

In spite of all these tendencies to grant increasing power to an oligarchy of privilege largely determined by wealth, American courts still uphold most of the provisions favoring individuals, including some often-unpopular ones involving gun control. In the balance, America remains a remarkably free society for individuals, and this continues to be supported by a large majority of the people aside from a small number of theocrats. Even more encouraging is the way intelligent conservatives are beginning to oppose the reactionary elements in the more conservative American political party. When some of the naive newcomers to politics wanted to close the government, hundreds of corporate executives wrote and said in effect, "Don't do that, you fools!" That was refreshing. It probably also reminded some of the extremists on the political right who really runs the Republican Party. The extremists and theocrats might find themselves politically marginalized again if they push too hard. So there are hopeful signs that the political winds may be shifting again, if not in a progressive direction, at least in a less reactionary one.

Democracy is defined as rule by the people. Enlightenment thinkers envisioned a world in which the conditions for democratic government could be achieved everywhere, even if most societies at that time had not achieved them. Beyond freedom of conscience and the basic human rights described above, additional conditions included peaceful tolerance for different political views and an educational level that goes beyond basic literacy to include adequate knowledge of the major issues facing society. Since no nation is completely isolated from others in today's world, this must now include understanding the major world problems affecting all nations.

Many contemporary nations are still not ready to elect successful democratic governments today. Upon gaining their political independence following World War II, many former European colonies in Africa adopted constitutions modeled after that of the United States. Some of those same nations have since succumbed to authoritarian rule, not due to any inherent inferiority of the people, but because of lack of preparation for operating a successful democracy under the conditions of their former colonial subjugation.

It is possible to set the bar too high in assessing the suitability of any society for democratic government. It is likely that no contemporary nation could meet all the conditions the original Enlightenment thinkers proposed for a democratic society. In a highly populated contemporary world in rapid transition, but still characterized by widespread ignorance, progress toward a distant goal is often the best one can hope for, even if that progress is frustratingly slow. If episodic democracy punctuated by authoritarian regressions at times of crisis is the best of which a society is capable, most secular thinkers would say this is still better than abandoning democracy altogether.

Beyond national democracy in the Enlightenment vision is a more unified humanity that has achieved an effective world government sufficiently strong to suppress major aggression by one member state against another. Religious parochialism, national chauvinism, and gross inequities in the global distribution of wealth all add to the problem of achieving world government. Nevertheless, promising trends toward a more unified world have emerged.

POLITICAL MOVES TOWARD WORLD GOVERNMENT

We must first note again the severity of several major problems facing humanity today. Ignorance, ethnicity, religious intolerance, and national chauvinism with its attendant militarism must all be largely overcome if any reasonable world order is to emerge. The current level of general education in the world is not encouraging, and the United States is no exception, all of which was noted in chapter 3 and will be discussed again in chapter 10. The role of ethnicity is only now being attacked by science, and the initial results are not encouraging, suggesting that ethnicity is a natural predisposition of people, which might have conferred an earlier evolutionary advantage. All major world religions span a large range from extreme parochialism to admirable tolerance, but each one has its rigid-thinking fundamentalists with whom reasonable compromise is difficult to achieve. The discussion of nationalism offered many examples of the widespread prevalence of national chauvinism and the attendant militarism. Until all these constraints on international cooperation have been overcome, it makes little sense to expect the emergence of a sovereign global authority that can solve great world problems. A sovereign world authority empowered to preserve peace among nations, promote universal human rights, and ensure global socioeconomic justice lies beyond any reasonable hope for today.

This does not mean that important initial steps have not already been taken toward world government, even if they are not recognized as such. A slow convergence of many actions currently taken or that are under way is taking place, operating seemingly but not entirely independently. The collective effect of these many actions is already laying the foundation for a future world government.[17] Just as we described some of the conditions for democracy in a nation, we now consider certain necessary conditions for effective world government.

Since the Enlightenment vision is global, it opposes all forms of national chauvinism as well as ideas of national exceptionalism. Consequently, if we embrace that global outlook, all forms of militarism that go beyond the defense of vital national interests immediately become suspect. Glory for its own sake becomes an unacceptable excuse for military adventures, and any use of a weapon more powerful than needed for defense against obvious aggressors becomes criminal. The presence of large national nuclear arsenals, or other classes of weapons with indiscriminate killing power, almost certainly makes the goal of effective world government unworkable. The nation possessing such an arsenal would have the means of blackmailing any international authority. Logic then suggests that no quick move toward world government is practicable as long as significant nuclear arsenals exist in the world.

On a more hopeful note, vital national interests of major nations today already involve treaties with foreign governments and binding economic agreements with other foreign organizations. While some of these agreements breed further problems, their purpose is to minimize friction in governmental relations and international trade. That the actual operation of organizations like the World Bank may be faulted on some counts does not mean that supporting a global economy is wrong, only that we are in the early stages of learning to make one work in a just manner. Learning to do that is another necessary condition for promoting a more harmonious world order.

Achieving a reasonable degree of socioeconomic justice globally will prove challenging. Different levels of economic development alone will make it extremely difficult for advanced nations to agree to a planetary community on terms that would likely be acceptable to the less well-developed countries. A progressive thinker might favor a world government modeled on a system similar to the one that produced the United States. In time, this might make sense. However, if we consider the possible consequences of a sovereign world government with effective powers of enforcement and a strong legislature based on nationhood *and population* (i.e., a world Senate and a House of Representatives), it is unlikely that any prosperous modern nation, and especially the contemporary United States, would approve such an arrangement today for fear of being exploited by a combination of populous poorer nations. It is difficult to imagine an effective world government unless large socioeconomic differences among major nations have first been overcome.

Yet a world government with effective powers of enforcement is the only sure means of addressing many otherwise-irresolvable global issues that are sure to arise or that have already arisen. Enforcing nuclear disarmament, once achieved; adjusting gross economic inequalities; and maintaining a healthy global environment have already been mentioned. Jared Diamond discusses ten major *current* global problems in his book *Collapse*.[18] As long as national sovereignty reigns, even those diplomats who grasp these problems will remain under pressure to protect their own nation's vital interests. Unless "win-win" solutions can be found, they will tend to seek actual advantage for those who sent them, as we have noted elsewhere. Near-term advantage can easily lead to long-term instability, but like the successful corporate executive who leaves with his large bonus, near-term success is a still a personal victory. The problem is systemic, or structural.

We are left today with negotiation of differences as the preferred method of resolving these issues, but the history of success in negotiation is spotty, especially if neither party has an interest in compromise. Not all problems are easily resolvable into win-win scenarios, especially when each side claims superior virtue. Such situations move nations ever closer to military confrontation. Can a vision that is necessarily global be realized in a world of sovereign nation-states when they are unwilling to compromise? The answer is obvious. In extreme cases, military action then becomes likely. In this case, Samuel P. Huntington's thesis in *The Clash of Civilizations* is confirmed, and his term "bandwagoning" to provide a defensive alliance against hostile nations may become necessary.

Even when international negotiations among sovereign nations proceed in good faith in a world lacking oversight of agreements that could affect other nations, problems can ultimately arise. An example of the danger of bilateral negotiations between nations that may ignore the rest of the world and ultimately produce unexpected future problems are the historic agreements forged between the United States and the leaders of Saudi Arabia during and after World War II. These agreements provided access by large American oil companies to the world's largest known oil reserves. I have no reason to think either side had malicious intent in producing them. When these agreements were made, there was little criticism of them on either side. Liberal contemporary Americans with limited historical knowledge may not realize that companies with the skills to extract oil, an often-dangerous activity, were usually considered war-winning heroes at that time, not environmental villains. To the Saud family, these agreements guaranteed a flow of American dollars sure to enhance their power and their nation's standing in the world. Almost everyone was pleased. Then, gradually in stages, came the realization that American dependency on foreign oil was itself becoming a problem, followed by scientific demonstration of the harmful effects of too much carbon dioxide in the atmosphere that only a few scientists fully appreciated when the original agreements were signed.

This example is not offered as a criticism of what was probably a good idea at the time but as a caution that sovereign nations—and unregulated multinational corporations—can and do look out for themselves, as they must, while the *long-term* consequences for both the sovereign partners and the rest of the world could prove large and not always beneficial. Whether future world government can exercise intelligent oversight on major international agreements that could affect everyone depends on how well the consequences can be predicted ahead of time. In the case of the impact of carbon dioxide in the atmosphere, that possibility was not widely recognized in the first part of the twentieth century. It has been now.

The point is not to identify heroes and villains, which is arguably problematic in all but extreme cases. We must note the fickleness of historical circumstances and contingencies. What may have been a good idea once may prove otherwise later, but by then vested interests and established ways of doing things have developed a powerful life of their own, making desirable course corrections extremely difficult to achieve. An effective world authority could speed up reaching rational global solutions in the future if our descendants can learn how to make one work. That alone will probably require more reliable knowledge of ourselves we do not yet have.

Perhaps there is one way that could lead to world government more quickly, in defiance of all the above predictions. Few would wish for it. A worldwide cataclysm resulting from global warfare or environmental collapse might lead the desperate survivors to produce one. The survivors would have good reason to fear the kind of government that such a catastrophe could produce.

While that dystopian scenario seems very unlikely, the problems identified by Jared Diamond and others will not be solved easily. Ignorance of how serious some of these problems could become will likely ensure a crawl toward solutions. Likewise, the move to world government is likely to be slow but is almost certain to occur in the future. Eventually realizing the universal humanistic goals of the Enlightenment seems very likely, but not in the lifetime of anyone living today. Nevertheless, it is useful to review further the arguments for achieving a more humanistic planetary society. The following chapter does that.

CHAPTER 9

PLANETARY HUMANISM

THE ETHICAL BASIS FOR PLANETARY HUMANISM

Planetary humanism is best described in the Neo-Humanist Statement issued in 2010 by philosopher Paul Kurtz, which is available on the website of the Institute for Science and Human Values (ISHV) at http://www.instituteforscienceandhumanvalues.net. This statement includes a summary of the ethics of secular humanism with special emphasis on their global application. It represents a modern restatement of the basic ethics of the original Enlightenment, which emphasized achieving a high quality of secular life for people everywhere. In contrast to the concerns of progressive Western thinkers during the eighteenth century when the Western hemisphere and Africa were still sparsely populated and many areas were unexplored by Europeans, the world today is densely populated and the surface area is well explored. In addition, modern technology has created global environmental and ecological problems that were serious only in local regions during the eighteenth century. Neither did earlier military technologies known to the original Enlightenment thinkers offer the possibility of destroying civilization as we know it, in contrast to the situation today. The Neo-Humanist Statement was written in recognition of these changed conditions.

We have noted several times that the goals of the Enlightenment are to be achieved through the application of science and reason. A simple demonstration in chapter 4 argued that the only alternative to this approach involves some degree of faith, since by definition science encompasses all reliable evidence and reason in its approach. Nevertheless, the guiding principles of the Enlightenment were and remain rooted in humanistic ethics. Since humanistic ethics are universal, encompassing all humankind, Enlightenment ethics have always been global, or planetary.

Some people dismiss these visionary statements as foolishly impractical compared to detailed plans for achieving well-defined particular goals. They tend to be critical of placing so much emphasis on broad principles. These critics have a point when plans for particular projects are needed. Nevertheless, they often err in underestimating the motivating power of appealing universal principles that inspire people to act, from which the details can then be, and often are, worked out with enthusiasm. In the absence of an inspiring vision for their lives, many people become bored or depressed. It has long been one of the functions of religion to provide an inspiring vision to compensate for life's many hardships and inevitable personal mortality. It is easy to understand why so many people living impoverished lives cling to religious superstitions, even in the absence of reliable evidence for the claims of their faith. Once more we note the need for a vibrant secular vision, if nontheists expect to compete with, much less replace, supernatural religion in the minds of people ill prepared to confront life's challenges with stoic acceptance.

The original Enlightenment thinkers realized that only by advancing the well-being of people in general was it likely that large numbers of them would abandon blind faith and superstition for a more rational way of life. They also recognized that the ethical principle of universal human betterment mandated a global solution. If one part of the world enjoyed prosperity while other regions were sunk in poverty, the situation would be inherently unstable. We see how true this is today, in an era of rapid global communications. Thus these basic goals for human betterment must not be narrowly nationalistic. The solution must be global.

THE PRAGMATIC BASIS
FOR PLANETARY HUMANISM

The broad humanistic vision just described has great value to a people for inspiring them to action. Nevertheless, while the vision itself is the right place to start, the "devil" still lies in the details. In this section, some examples of pragmatic approaches to practical questions are discussed. We note that answers to these questions involve the application of broad ethical principles.

American philosopher Austin Dacey has addressed the question of how a secular person in a religion-saturated country like the United States, or anywhere, might conduct himself in the public square. In his book *The Secular Conscience*, Dacey argues that the state should welcome the public expression of all possible views short of inciting immediate danger.[1] Reminiscent of John Stuart Mill's essay *On Liberty*, the book makes the case that a society will benefit more from unrestricted freedom of speech than from suppression of individual opinion.[2] Dacey would not limit the right of Catholics to publicly argue for legal restrictions on abortion, nor would he limit those who favor no restrictions at all. The principle of free speech in dialogue and writing is embraced by humanists universally.

Another example of what might be called *applied humanism* is a well-known study of the basis for a society's laws, though it can be seen as a culmination of a long line of democratic thought that developed with the Enlightenment. John Rawls's idea of a just society is based on the surprisingly simple idea of fairness.[3] This sounds trivial at first thought, for what is justice but fairness to all members of a society that aspires to equal justice under law? However, this still leaves the question of how a society's laws are to be achieved. Rawls devises a method (described in chapter 8) that begins by placing all members of a society on an equal footing. While this has yet to culminate in a successful world government, forming a more effective planetary society has already begun. Many international treaties and international organizations, most notably the United Nations, have initiated this process.

Some of the ways in which humanity is moving toward one global society are discussed by former ambassador to Nepal Carl Coon in his book *One Planet, One People: Beyond "Us vs. Them."*[4] However, Coon also reminds us in *Culture Wars and the Global Village* that people have lived in contentious groups too long for intergroup hostility to be ignored, as we move toward a single global village.[5] This problem is unlikely to be solved quickly, and there is reason to move in a number of sequential smaller steps. Today, the wealthier nations would probably resist rapid movement toward world government, for the reason given in chapter 8.

Going into more detail in applying ethical principles and modern science to specific cases is Ronald Lindsay's *Future Bioethics*, a concise, well-argued summary of the ethical and scientific bases for many still-controversial developments that offer people promise for better lives.[6] Examples include stem-cell research, compassionate physician-assisted end-of-life laws, genetically modified crops, and improving innate human faculties such as intelligence. These developments all involve the application of leading-edge research in the biological sciences. While some of them remain highly controversial, progressive thinkers today are tackling increasingly complex societal problems from the perspective of modern science, while guided by the broad humanistic outlook that characterized the original Enlightenment.

In *Forbidden Fruit: The Ethics of Secular Humanism*, Paul Kurtz has demonstrated that ethics and morality need not be a highly abstract subject when presented in language accessible to any reasonable person.[7] Kurtz defines the common moral decencies as those moral prescriptions that have arisen from human historical experience, needing no divine lawgiver for their origin. While supported by many academic studies of ethics, these straightforward moral guidelines are familiar to all decent people. Many of them have been incorporated into the moral codes of all the world's major religions. They constitute an easily understood practical guide for everyday conduct.

The Washington Area Secular Humanists used Kurtz's book to initiate a year-long discussion of ethics and morality. We generated our own list of twelve guidelines for moral living, based on humanistic ethics. Taken together, we concluded that these guidelines represented a set of values that would, if followed, promote a humanistic democratic society. Our results were subsequently published in *Developing Sanity in Human Affairs.*[8] Kurtz's prescriptions and those resulting from our discussions adopted a different approach—ours being very "grassroots"—but the two approaches arrived at the same conclusions. Both also emphasized that humanistic ethics and morals are the product of human historical experience.

While there is much commonality among all societies on basic ethical principles, there are also notable and occasionally disturbing exceptions. To understand these exceptions, we must often turn to the impact of a people's religion. When different religions make different dogmatic claims, adherence to a common set of basic humanistic values becomes increasingly difficult. Obviously, this impedes the move toward one stable world community.

One example is given by certain contemporary teachings of the Wahabi sect of Islam. This sect is not only growing in strength today; it is also one of the most rigidly dogmatic in its interpretation of the Prophet's message.[9] The Wahabi interpretation of sharia and jihad differs from that of the more accommodating Islamic sects with their tradition of "agreement." In this latter case, the leaders can interpret the original Qur'an more liberally to adapt to contemporary conditions of life. Perhaps the impact of education and the recent Arab Spring will modify the more dogmatic views of the Wahabi leaders.

We can hope for a similar development in the United States. Another example of rigid dogmatism is Protestant fundamentalism in America. There are many variations of these sects, and not all are theocratic, but some are. Among the latter are those that would radically alter the American Constitution and force a theocracy on the country. While generally less rigid in adherence to ancient texts, traditional papal Catholicism seems reluctant to retract its insistence that there is no salvation outside of the Catholic Church, though some papal statements have suggested otherwise. Ultraorthodox Jews may claim an eternal right to all of Israel or even a "greater Israel," based their interpretation of an alleged ancient promise from God.

The conflicts between, and even within, the more dogmatic religious

sects can be intense and can occasionally turn violent. The leaders and followers of these sects who resist all compromise may constitute the greatest
clear and present danger to further human progress today. Yet, probably
only a small fraction of the people adhering to these sects are so uncompromising, with even fewer being truly dangerous. At the same time, we
might recall an important lesson from history. A small, organized group of
fanatics can have a large impact on any society unwilling to oppose them.

THE PROSPECTS FOR PLANETARY HUMANISM

Predicting the future of our descendants in detail is impossible, so we can
only address some of the major progressive developments that will advance
humanity toward a stable global human community. Since major changes
in society do not normally occur quickly, trends are critically important.
There are a number of world trends that would certainly advance traditional
Enlightenment goals. These include trends toward a more just and equitable
distribution of the world's resources and products; toward establishing a
sustainable use of the earth's resources to preserve a healthy environment
and ecosystem; toward establishing equal rights, opportunities, and respect
for women; toward preventing wars between nations through development
of a sufficiently effective world government; and toward achieving at least
tolerance for different philosophical and nonviolent religious positions,
while preventing both political and/or religious fanaticism from interfering
with these achievements.

However, it must be acknowledged that several current world conditions and powerful contemporary attitudes and interests stand in the
way of these achievements. The previous section named one: religious
intolerance and fanaticism. Others include widespread poverty; chauvinistic nationalism in powerful nations; entrenched economic vested interests; over-exploitation of the earth's resources with inevitable damage to
the environment and the ecosystem; and serious human-rights abuses,
including sexism, which still affects half the world's population to varying
degrees. Traditional attitudes can offer individuals and societies near-

term stability, but they can also be major obstacles to change, when reason argues that change is necessary. Complicating the problem, we now know that the rate at which people can learn, individually and collectively, sets limits that affect how rapidly individuals and societies can adapt. We are not infinitely malleable.

Thus achieving the goals of planetary humanism is unlikely to occur nearly as quickly as many of the original Enlightenment thinkers hoped. The knowledge and experience we have gained since the eighteenth century confirm this beyond a reasonable doubt. Yet the goal remains compelling from a rational perspective. Each individual, independent of his or her status, is placed behind a hypothetical curtain and asked to decide what each thinks the laws should be. When all members have done this, groups are assembled to discuss and argue among themselves what form the laws should take. When each group has deliberated and come to some agreement, successively larger groups can take the results of the smaller group's deliberations and carry the process further. The process bears a rough relationship to the way laws and treaties are eventually formed and is as democratic as can be expected in a world where individual native talents are not distributed equally, but the dignity of all persons is recognized from the start.

PART 4

REASSESSING THE

ENLIGHTENMENT

CHAPTER 10

OVERCOMING CRIPPLING IGNORANCE

LIVING WITH IGNORANCE TODAY

So far, this book has made a case for both significant achievements and serious shortcomings in the effort of the past few centuries to produce a more satisfying world for people everywhere. Earlier chapters noted the almost-universal agreement on the amazing scientific and technical progress of the recent past but a wide difference of opinion on whether this has led to substantial overall societal progress as well, especially in the area of socioeconomic justice.

We have already noted the optimistic belief in endless progress of many thinkers during the eighteenth and nineteenth centuries, particularly in the West, and how this has changed to a more sober-minded assessment of near-term possibilities by equally thoughtful people today. We have also noted the reluctance of many institutional beneficiaries of this scientific and technological progress to admit that these advances have introduced disturbing harmful effects as well. Typically, their arguments make valid statements about the impressive progress, while ignoring the corresponding detrimental effects that have been well documented by others. Another argument of the techno-optimists is to claim that the solution lies with more of the same technology that created the problems in the first place. While all arguments are worth considering, the latter are becoming

175

increasingly unconvincing to a growing number of scientifically literate people today.

One reason the extreme optimism of the Age of Reason has proved premature is the faith many Enlightenment thinkers placed in the role of reliable evidence and reason *alone* in promoting human progress. Even today, it is possible to find highly educated people in the academic world who cannot grasp why there are intelligent people who refuse to submit their opinions to evidence and reason. A greater familiarity with the better psychologists among philosophers would have suggested a likely explanation. To assess how reason might manage passion, it is first necessary to understand the nature and role of passion. In reassessing the Enlightenment, one of the first conclusions we can draw is that a more subtle scientific understanding of the human mind/brain is needed than was available to some of the more optimistic thinkers . . . then and now. This conclusion falls directly out of the discussion in chapter 5, where neuroscience is briefly reviewed.

There is no longer any doubt in the minds of many thinking people that widespread ignorance remains a major obstacle to rapid societal progress. The nature of this ignorance is discussed in the next section of this chapter. However, that ignorance is pervasive is demonstrated by the frequent failure of many schemes developed by conscientious thinkers for the rapid improvement of society. Many of these schemes have a history of spectacular failure. We are therefore confronted with the question of how to proceed when the kind of knowledge science can provide in principle— but has yet to provide in practice—is lacking.

In the current absence of a reliable, mature neuroscience, we still need to rely on knowledge that has been acquired of human propensities and behaviors that have been observed and studied through the ages. Certainly this knowledge is not trivial, even if it does not yet meet the high standards of science. One of the strongest arguments for establishing tolerance among secular thinkers who reject all religious dogmas and the tolerant religious is that some of this knowledge based on experience alone is still transmitted to many people through religion. This does not require the secular thinker to accept the alleged dogmatic bases, nor does it require him to cooperate with religious reactionaries, many of whose ideas are demonstrably dated or even absurd. It does acknowledge that until we have more reliable scientific knowledge of the physical basis of our mental functioning, we still need to rely on experience in those areas where that knowledge is lacking.

This need to rely on experience alone, when necessary, has consequences. It makes it possible for people to select from a wide range of attitudes toward the future, and to *feel justified* in their widely differing views. While this book argues that the reliable knowledge we already have strongly favors the progressive position on most societal issues, intelligent conservatives play the useful role of criticizing the more poorly thought-out schemes that are sometimes advocated "in the name of the people." Some of these schemes have proved to be open invitations to tyranny.

We can ask if there is any other source of insight beyond collective wisdom that can aid us where science is still inadequate. There is, and it is the field of ethics, a major subdivision of philosophy. Because of the importance of ethics, I find it surprising that modern education seems to largely overlook philosophy as a field worthy of serious study. A doctorate in philosophy can be an invitation to unemployment in the United States today, yet, with the possible exception of science, there is no field more attuned to critical thinking. Philosophy may have yielded largely to science in the area of epistemology and may be too esoteric for pragmatic applications in the field of metaphysics. However, philosophy continues to be of critical importance in the field of ethics, where the naturalistic fallacy still reigns unchallenged by serious thought, at least based on what is known today.

We have noted several times that the goals of the Enlightenment were profoundly ethical when viewed from a secular humanistic perspective. Even the pursuit of science and the reliance on reason can be viewed as instrumental values, or as the recommended means to realize these goals. This places the field of ethics at the forefront of deciding what we should do in addressing major societal issues. A current example in America is the political battle between those aware of the reality and seriousness of human-driven climate change and those who for parochial reasons attempt to deny human causation. Recently, an effort was made by scientists to educate Congress and the American public on the scientific basis for their almost-unanimous scientific conclusion on human causation of global warming. For many reasons, this effort failed. That was followed by a modified but equally unsuccessful effort to emphasize the disastrous long-term global societal consequences to these same people. That effort also failed, largely due to a well-funded lobbying effort by energy industries. Now, educated Americans are finally beginning to recognize that the issue is ultimately an ethical one, something recognized by Al Gore years ago but now taken up by scholarly studies that should become better known.[1]

I conclude from this that a more ethical approach to great world problems is badly needed, especially in my own country today. Americans are famous not only for quickly reaching pragmatic solutions to major problems but also for an obsession with material comforts and the famous/infamous "bottom line." There are obvious advantages to this pragmatism. Few of us would like to seriously harm or radically change the powerful economic engine that has provided a high standard of living to our large population. However, this does mean that more effective regulation and some genuine course changes are not needed when they clearly are. It will be interesting to see how these changes come about, as they surely will, sooner or later.

Thus I conclude that while we remain ignorant of many things bearing on major world problems today, we are far from helpless. Science *has* given us much relevant reliable knowledge of the physical problems facing us, and some knowledge of the biologically based problems as well. Tolerant scholarly religion has accumulated a vast amount of historical knowledge that should not be ignored, even by those of us who find all God-based dogmas unsupported by evidence. The field of ethics has been richly mined by philosophers, many of them secular, and has much to offer if more thinking people would take it seriously.

Nevertheless, while we have certainly proceeded far beyond the days of Socrates and the remarkable Greeks in many areas, our reliable scientific knowledge of even the individual person remains relatively primitive. When we ask what happens when large numbers of people in a society interact to solve their collective problems, the complexity of the first problem is multiplied further. We cannot escape the conclusion that we remain ignorant of many things that would be useful to know well, if we are to solve these major societal problems in an optimum manner. From this I conclude that the greatest problem still facing the human race remains what it has always been ... *widespread ignorance.*

If ignorance is indeed our only one true "enemy," then we need to grasp what ignorance is. *Ignorance* is often defined as simply lack of knowledge. However, that alone is clearly inadequate. There is a great difference between being aware of one's ignorance as opposed to being unaware of it. There is also the fuzzy area of superstitions that might not contradict reliable knowledge, but lack confirming evidence. I will not review these subtle points here. Instead, I will define and discuss here two easy-to-understand and separable kinds of ignorance that are relevant to this book.

CRIPPLING IGNORANCE OF TWO KINDS

I define *ignorance of the first kind* as ignorance of knowledge that is already available. While there are many examples of such knowledge, the results of natural science clearly represent this kind of knowledge best. I argued in chapter 4 why scientific knowledge is the most reliable kind we can hope to gain as long as we restrict ourselves to the natural order. Of course, if there is no other order, in particular "the supernatural," then scientific knowledge becomes the most reliable kind of knowledge we can attain of everything in existence, including ultimately even the fantasies of the human imagination. It would seem straightforward to argue that as long as we do not insist that the natural order is the only order—even if that is our personal position—science education could proceed without hindrance from anyone, given the obvious importance of science to the modern world and to our descendants' future. Unfortunately, attaining this in our educational systems has proved to be far from simple. People addicted to religions of the book, which insist on the inerrancy of ancient prescientific texts, have often stoutly resisted all science that conflicts with any statement from these texts. Thus, education has failed in many parts of the world to overcome ignorance of what is already reliably known. This first kind of ignorance remains widespread in the United States today. Inadequately addressed, it will have serious future personal and societal consequences by inhibiting medical research if blocked by the nonsense of "creation science" and by delaying rational approaches to climate change if blocked by the so-called sound science of the climate-change contrarians.[2]

In fact, this opposition to reliable scientific knowledge has had an even more destructive consequence than its denial of well-established scientific results. It has led many of the opponents of established scientific theories such as evolution to attack science itself. That this opposition to science, which is often poorly understood by these same individuals, could eventually undermine the modern economy many of them ultimately depend on is overlooked by them entirely. Lacking proper education in the nature of science, many of these individuals also welcome the antiscientific position advocated by some scientifically ignorant post-modernist and deconstructionist intellectuals. Many aficionados of these disciplines—but few trained scientists—aver there is no superior way of knowing anything, which is appropriately described as "higher order superstition" by authors Paul R. Gross and Norman Levitt.[3] Susan Jacoby's Age of American Unreason provides another survey of antiscientific irrationality in the contemporary United States, especially in her chapter appropriately titled "Junk Thought," a stand-in for the pseudoscience that is often encountered in political advertisements and even in congressional briefings.[4]

Unfortunately, junk thought seems to be growing in many other parts of the world today. It is a major issue in many contemporary Islamic countries, where creationism has become popular among some of the more radical Islamic clergy. There could be few better examples of competing reactionary religions than aggressive Protestant fundamentalism in the United States and equally aggressive Islamic fundamentalism in these other countries. Yet the two have much in common, psychologically, and share a rigid closed-mindedness to science when it conflicts with their ancient texts.

The problem of ignorance of the first kind is especially appalling, because it could be overcome today by a dedicated effort to properly educate people. Lacking the desire to do so is an ethical problem, and those actively responsible for this are likely to be judged severely in the future. Yet the problem of ignorance does not stop there. Another kind of ignorance must be considered if we are to assess the prospects and determine the means for achieving a more humanistic society. This is ignorance of what we will need to know to greatly improve our chances of solving many of the great world problems already discussed here, but that is not reliably known by anyone to date, especially in the realm of scientific knowledge. A particularly disturbing example reviewed in chapter 5 is the lack of an in-depth scientific theory of the human brain, the central organ in the human body that determines the decisions we make in our effort to solve these problems. The brain is the physical command center for what we do. Leaving the metaphysical problem of the mind versus the brain aside, we can truly say that under naturalistic conditions: "no brain; no mind." This example of our collective ignorance leads directly to the following definition of ignorance of the second kind.

I define *ignorance of the second kind* as ignorance of anything that has so far defied human efforts to acquire reliable knowledge. I will argue in what follows that ignorance of the second kind, along with widespread ignorance of the first kind, will make it difficult to solve many of these great world problems quickly, a disturbing prospect since some of these problems bode ill for humanity in the near term if unanswered by intelligent collective action.

Finally, to grasp the magnitude of the problem of ignorance, one more critical aspect of the challenge needs emphasis, the radical difference between education and training. Training in the proper use of words and numbers is an essential first step toward an educated mind, but it does not constitute education. Education necessarily involves learning how to think critically. This consists of learning how to ask good questions, how to identify a problem, how to pose a problem clearly, and how to recognize the difference between a nontrivial but potentially solvable problem and one that may lie well beyond one's current capabilities. It further involves knowing how to analyze a proposition, how to critically assess underlying assumptions (especially the hidden ones), and finally how to think logically toward a conclusion. All of this constitutes a huge difference between mastering only simple reading, writing, and arithmetic and learning how to push buttons on a computer.

This is well-known, or should be well-known, to professional educators, but it may not be fully grasped by others unless people are made aware of it. Simply stated, education consists of learning how to think straight and recognizing why this is both personally useful and important for others. The importance goes well beyond appreciating the role of science in a modern society. It goes to the heart of democracy itself, where lack of education leaves the ignorant easy prey to political propaganda. If the mass media could be dominated by a small number of skilled political propagandists, it would not be necessary to suppress democratic voting. It would then become easy for an unscrupulous minority to induce the great multitudes to vote in opposition to their own socioeconomic interests, a perfectly peaceful pseudodemocratic solution to a modern oligarch's dreams.

WHY QUALITY PUBLIC EDUCATION IS NEEDED

Education has obvious personal benefits. Many statistical studies show that the higher the level of education a person reaches, the larger the income that person is likely to command. Yet the value to the community at large is arguably even greater. The importance of quality public education to politics in a democracy cannot be emphasized enough. Jared Diamond may do his best to paint a hopeful picture for the near future in the final part of *Collapse*, but when one reads his list of ten major world problems that continue to defy solution today, it is hard to think his qualified optimism for the near future is entirely justified.[5] People need to know what these problems are.

In a similar vein, after offering a devastating critique of neoconservative—which some call neoliberal—multinational corporate capitalism, Naomi Klein ends *The Shock Doctrine* with a ray of hope.[6] She notes that the majority of people in the first countries subjected to the shock of an unregulated free market have partially recovered and are bringing the exploitative efforts under more effective control. It is possible these trends will continue and succeed elsewhere. The recent Occupy Wall Street movement in the United States bears watching as of this writing. People who do not know there is a still-unsolved structural problem with unregulated mature capitalism are poorly equipped to deal with it politically.

If actions are being taken by some to control corporate excesses when things run wild, there are others who are watching, and they hold the largest purses. The power of well-financed modern propaganda in the absence of adequate public education can be decisive. Even in an allegedly modern country like America, all a reactionary politician needs to do to win millions of votes is to wave the American flag and sprinkle the word *God* in his speeches. One might think certain Republican propagandists had read the diaries of Joseph Goebbels and taken a few tips from them. While adequate public education requires a few years to bear fruit, there may be no other remedy to turn this deplorable situation around. It is not surprising that political reactionaries are the first to oppose effective education, since they are the ones most likely to be exposed by it.

This raises the question of how to achieve effective public education, which is essential for applying people's critical faculties to civic problems. That public education is in a precarious state in America today is generally acknowledged. While education of a receptive minority for success in the world of advanced technologies and global corporate business is proceeding well, and those of us who benefited from that have enjoyed productive lives, practically every poll and every statistical assessment of socioeconomic well-being has shown that the plight of the majority in America has worsened. Clearly something is wrong, unless you think the life of the hypothetical average person should be subjected to paternalistic management through provision of the material basics for life, along with consoling secular or transcendental myths.

This patronizing view of "the people," once common in aristocratic circles, continues to be alive and well today. Only today, the former aristocracies of birth have changed into plutocratic financiers and managers of the corporate economic juggernaut, along with those who provide them with refined academic ideologies or who devise their propaganda. Nor are these the aristocrats of merit Jefferson spoke of, replacing those of birth.

Instead, the current plutocrats possess the one thing history shows is both most necessary for life today and most likely to be corrupting. The modern aristocracy is based on money, far in excess of what is needed for a life of comfort and reasonable philanthropy. Educated professionals in scientific and technical fields who serve this system are well provided for, but the majority of people are increasingly reduced in well-being. Whether called *trickle-down economics, supply-side economics,* or *the rising tide that will raise all boats,* this system, unchallenged and unregulated, has so far demonstrated only its ability to make a small minority very rich. An ignorant population lacks the means of assessing this situation and finds it difficult to correct.

That adequate public education can reduce the lock that money alone can easily place on the global economy—and through it, on global politics—is self-evident. But since education is not the same thing as training, skepticism is needed in assessing educational initiatives, particularly "No Child Left Behind." In this program, traditional kinds of protoeducation (reading, writing, and arithmetic) are honored, but critical thinking may or may not be adequately taught, depending on the instructor and the state in which the curriculum is developed. Education beyond basic skills may not be desired by those who see "the people" as mainly a source of workers in the corporate economy, as mothers for the next generation and as soldiers for an occasional battlefield as the alleged need arises. Quality education is then restricted to an elite group of financiers, managers, military officers, academics, and researchers, especially those who support research with direct bearing on the corporate economy. This is a traditional notion, and one that is still very much with us. Historically, there are many reasons why this arrangement may have been optimum for an earlier time. However, if "the people" are to be more than a workforce to be managed by an aristocracy of wealth, this arrangement has serious defects.

In her claim that when a people recover from socioeconomic shocks they often take matters into their own hands and reverse many of the egregious measures taken against them, Klein is, in effect, arguing for "the wisdom of the people." This resonates with John Stuart Mill's arguments presented in *On Liberty* for the power of free speech, a free press, and so on.[7] However, everything remains contingent on the assumption that the people are actually able to figure things out for themselves. Just as Adam Smith's ideas on unregulated free-market capitalism were revolutionary for his time and are arguably anachronistic for ours, we must note that the society of Mill's day was also simpler than the complex global society of today. Education needs to improve if the people are to effectively play the role that Mill envisioned for them. Otherwise, authoritarian government becomes inevitable, even if it retains democratic trappings.

If the power of the dollar and political propaganda seems exaggerated in these paragraphs, the following considerations suggest otherwise. A large American energy corporation recently made forty billion dollars in profit in one year. If further frustrated by climate scientists, might not this corporation offer a mere billion to Madison Avenue for more television sound bites to suppress the claims of anthropogenically driven climate change that climate science has revealed? Are people who need no further assurance that a politician deserves their vote than to see him wave a flag and refer to a deity capable of analyzing even basic contemporary issues?

Once again, Mill's confidence in the people (and perhaps Naomi Klein's) seems up for another test. *Do the people realize that they too are on trial here; and that if, in its attempt to solve major problems, democracy fails in a way that hurts them, the people must accept some of the responsibility?* Democracy was discussed earlier as a complex system that could not guarantee good government, even if it is our best antidote against tyranny. This involves an implicit assumption that the people are not easily fooled. Perhaps the people need more than just a better education. Perhaps after living in a country that has lived "high on the hog" for over half a century, the people once more need to be challenged. Perhaps this will happen naturally. We have all evolved to be problem solvers.

We noted that ignorance is of two kinds. The first kind was defined as ignorance of reliable knowledge already available if a society has the will to deliver it through a public educational system. (Would any reader of this book trust an entirely private or parochial one?) The second kind is ignorance of knowledge that is not yet available. This second kind could still be essential for solving many current world problems in a satisfactory way, but widespread ignorance of the first kind is particularly deplorable. Examples of this first kind include knowledge of what science is, what some of its more reliable results are, and of how a democratic civic process works and of what it needs to work; that is, public concern and participation. Two examples of the second kind are a reliable theory of how the human mind/brain works with the predictive power always associated with a mature science, and a reliable theory of economics that incorporates the irrational human factor. Note that getting such a theory of economics requires progress on the first of these two problems. Can anyone today create a mathematical model that can predict the future behavior of the stock market, as opposed to predicting how best to react to changes in the economy?

The kind of education referred to here is that which addresses ignorance of the first kind. If we could go *just that far*, we would be far beyond where we stand today. It would be even better if most people understood the nature of the still-unsolved problems, but that may lie beyond current public educational resources. Just taking the first steps beyond reading, writing, and arithmetic to an understanding of what science is, what its more well-established results are, and how a democratic political system works would give people today a much better chance for solving some of these compelling problems that have emerged in the contemporary world. Most disturbing is that many on the reactionary side of the political and associated religious communities may actually be opposed to such effective public education.

Fortunately, there may be one feature of contemporary society that can leverage what even excellent formal education alone cannot achieve. Many contemporary thinkers have noted the near impossibility of regulating the Internet, something that has given even the highly sophisticated Chinese bureaucrats nightmares. Even if public education in America and elsewhere is demonstrably weak in several critical areas, literate and energetic young people have become quite skilled in Internet communications. Adding this to the healthy skepticism of youth, and a natural desire on the part of a growing number of smart young women joining these young men to displace stodgy older men in positions of societal power, the corporate world itself could change in a way the current captains can hardly imagine.

OVERCOMING CRIPPLING IGNORANCE BEFORE THE DELUGE

This section will give one example to demonstrate why overcoming crippling ignorance is critically important to human welfare and progress today. I have chosen an example in a field for which I have some professional knowledge, thanks to another in which I worked for most of my professional life. The first field is climate science; the second is solar physics. As with other cases in this book, my examples will be drawn from experiences in my own country, because I know the situation here best; but like many other problems noted here, this one is global in its impact.

It is first important to reiterate a few things about science itself. A proper public educational system would have already prepared the American public to assess science-based issues. Unfortunately, we have not reached that stage yet. The United States educates some of the world's best research scientists and provides some of the most advanced facilities for research. Yet the American public's understanding of science as a whole is relatively poor, and in some American subcultures, it is appalling.[8] This naturally makes people vulnerable to political manipulation through what can only be called propaganda.

To be able to assess science-based problems, the public needs to understand what science is and how the process of science works. In chapter 4, science is described as a process through which we get the most reliable knowledge possible of how the physical world of nature works. Most people probably understand that much, but many do not grasp that the process is more important than the results. So when a person in the political realm, especially a scientist, makes a claim in the name of science, there is a tendency for many people to accept it uncritically. Using climate change as an example, a few people who have done science have made highly dubious claims in the name of science to counter the efforts of a large climate-science community to discredit legitimate climate science, sometimes calling their own version "sound science."[9] This is a clever trick, applying subtle psychology of how people, especially ignorant people, react to language. In fact, only a very small number of actual climate scientists remain skeptical of the following conclusions: (1) Global climate change is occurring. (2) This change is proving to be far more harmful than beneficial, and the harmful effects are almost sure to get worse. (3) Global climate change is due primarily to extremely well-documented global warming. (4) This global warming is due primarily to the increase of greenhouse gases in the earth's atmosphere, an increase for which human industrial activity and transportation are largely responsible. All these statements can be defended by an enormous body of research.[10] However, the American public's limited understanding of science allows a small number of climate-change contrarians to block government efforts to address the problems of global warming.

The way in which the climate-change contrarians have attempted to deny human causation is instructive. First, many contrarians attribute climate change to the sun. There is no credible evidence that the sun is primarily responsible for the rapid global warming of the last three decades. Solar physicists have looked for solar causation and have failed to find it.[11] This demolishes the false claim that scientists are using global warming to bring more funding into their respective fields. Since solar causation would have substantially increased funding for solar studies, the solar scientists were clearly not following this business model in reporting the results of their research.

The final observation on attempts to discredit climate change concerns "Climategate." Anyone who believes that these results were only coincidentally released before the 2009 Copenhagen meeting on global warming and climate change is very naive. It is reminiscent of the boy who is discovered with his hand stuck in a cookie jar, a situation some describe as "accidentally on purpose" when the boy says, "Why, I didn't know this was the cookie jar!" I knew the Climategate critique was spurious, and I realized what a desperate state of mind the climate contrarians were in, determined to prevent action to mitigate global warming. They had sought an amazing variety of natural mechanisms that might explain global warming and had failed to find even one. Climategate gave them the best hope of resisting political action with the Copenhagen meeting staring them in the face. My published comment was "Is this really the best you can do?"[12]

That said, here is the sobering news. Climategate worked. The Copenhagen meeting accomplished little of note. Generously funded propaganda for public "education," massive support for lobbying Congress to block ameliorative legislation, and courting a few formerly distinguished scientists for support has worked. As of this writing (late 2011), little has been done by the American national government to address climate change. If effectiveness in addressing the problem of global warming is the criterion for judging a winner, the American scientific community, and hence the public, have so far lost on this particular issue. That is not for lack of trying by the climate scientists, but an educated public could have made a big difference. Science will go on. It is the people of the world who will suffer the harmful effects of climate change, but most of them in this country don't seem to know it yet. What you don't know *can* hurt you.

CHAPTER 11

REASSESSING THE ENLIGHTENMENT TODAY

THE INERTIA OF TRADITION

Traditions arise by surviving the test of previous human experimentation. They are not to be taken lightly. As long as the conditions that applied when the traditions were established continue, it is often better to accept the collective wisdom they incorporate than to take an unreflective move in a radically new direction. Anthropologist Lionel Tiger has noted that resurrection myths made sense to earlier people who lived as illiterate farmers and shepherds.[1] They personally experienced the apparent seasonal death and resurrection of life in the natural world they lived in daily.

The problem begins when familiar conditions undergo rapid change. Some people, especially the more curious and adventuresome of the young, adapt to the new conditions quickly. Older people, especially those who have mastered the challenges presented by the old ways, tend to be more conservative and sometimes stoutly oppose change. In a similar fashion, mature societies and corporate structures often exhibit the same behavior once success has been achieved, overlooking that it was radical innovation that created these successful societies and organizations in the first place.

American history offers one particularly good example of this. The American Revolution may have been the most moderate of the three attempts at progressive change in modern Western history. However, it was not only the first; it was also the most successful. In the effort to promote complete institutional separation of church and state—still never achieved permanently by any large society to date—the American Revolution attempted to implement one of the most revolutionary ideas in history. Today, we see American political conservatives and reactionaries using legal and extralegal means to overturn or subvert many of the barriers between the institutions of religion and those of government.

In the economic realm, the American steel industry and the American aerospace industry illustrate a similar trend toward conservatism with maturity. The first eventually experienced major economic failure after impressive initial success. The second is now searching for markets after spectacular expansion. A current industry that bears watching is the complex of American energy corporations. These giants seem torn between their desire to continue innovating, something they have done well in the past, and their wish to silence pesky scientists who have demonstrated that burning coal and oil is already changing the planetary climate. Eventually needed change comes, but seldom as quickly as a more rational policy might dictate.

Thus the inertia of established ways will always be a major obstacle to progressive change, even when the evidence greatly favors it. One of the advantages of democracy in contemporary societies is the provision of a relatively peaceful mechanism for managing needed change when it is likely that some will strongly favor the changes while others oppose them. Lacking such a mechanism and a strong tradition of law for settling the issues, history demonstrates that violence often occurs when major institutional interests are at stake. That the more civil result may be a compromise that satisfies neither party is generally regarded as better than open conflict, but this requires a willingness of both parties to back away from extreme positions, even if one position is backed by science. The frequently mentioned issue of climate change again comes to mind today. Most people would probably agree that the suffering likely to result from an additional rise of one or two degrees Celsius, while disastrous for certain parts of the globe, is probably less than that which would result from another global war. On the other hand, even studies done by the American military take seriously the possibility that climate change along with resource exhaustion could be important factors in precipitating a major war.

Nevertheless, the example of climate change supports one major conclusion in the assessment in this book. This conclusion does not anticipate rapid progress in solving several major world problems quickly. We have already noted a number of disturbing changes in the earth's ability to sustain current population growth with present rates of resource exploitation and the use of contemporary technologies. Yet the political pressures for addressing these problems continue to be resisted by those favoring business as usual, often presented with reassuring statements that new technologies will be available soon enough to avert the more serious prognoses of disaster. This requires a sobering look at the road ahead over the next century before expanding on the more distant future. Partly because of the inertia of established ways, that road is not likely to be smooth.

THE ROUGH ROAD AHEAD

While entire chapters in this book have been devoted to religion and politics, the subject of economics has so far received little attention. Yet when considering the time people devote to their struggle to survive and to thrive, economic activity, broadly defined, consumes the greatest amount of time, especially if the related activities of caring for the young and the very old are factored in. Economic activity provides for the physical foundation of human life. While "man" may not live by bread alone, people certainly need bread first. Many treatises on war note how much economic motives underlie military aggression, even when disguised by other motives. Even religious wars often conceal economic motives, for which religion is sometimes unfairly blamed. Studies of people suffering from starvation note how quickly other interests, particularly interest in sex, disappear when people are hungry. It does little good to reproduce if you cannot feed yourself. Nature has trumped culture in determining personal priorities.

The most successful economic model for maximum production to

date is capitalism. Once the current Chinese government permitted entrepreneurial economic activity to thrive, the Chinese economy developed
explosively, and that growth continues today (2011). Various forms of
socialism have been tried and have so far largely failed, often from bureaucratic strangulation. At the same time, it is generally conceded that a major
economic problem has arisen as a result of the success of the modern
mature multinational corporation. That problem has created conditions in
the contemporary world that run counter to the humanistic goal of a just
socioeconomic order, which necessarily entails a just—though in general
not an equal—distribution of wealth among people. This situation is one
manifestation of the already-discussed structural problem of inadequately
regulated mature capitalism. By its very nature, this system exhibits a well-
documented historical tendency to produce first monopoly, then some
form of imperialism, and finally war, as noted in chapter 2. Yet the adaptive ability and productive potential of this same contemporary economic
engine is extremely impressive. Not all global problems should be blamed
on inadequately regulated capitalism, but few other human developments
have had such a strong impact on contemporary life today. There is irony in
this situation, for capitalism has *under certain conditions* brought real benefits to hundreds of millions whose economic standard of living has been
raised by it. Yet there is equally little doubt that this same productive engine
has often run wildly out of control, creating either huge material inequalities between the few and the many, and/or leading to highly destructive wars. Our inability to solve this problem may be the best example of
humanity's failure to date to successfully manage the consequences of the
rise of modern science and the Industrial Revolution, events commonly
associated with the Enlightenment in the eighteenth century.

Common sense suggests that the solution to this problem is adequate
regulation by government, the only entity capable of regulating the productive capacity of a people whose economic welfare depends on its proper
operation. Determining what amount of regulation is appropriate between
that inadequate to prevent destructive consequences and one that stifles
innovation is a problem requiring experience and expertise. However,
only a rational mind is needed to expose an absurd view that seems to

have taken root in the United States since the collapse of the overregulated former Soviet empire. Some who point to this example seem to have concluded that government regulation is *always* the problem, so the solution is simply to eliminate all regulation. This is tantamount to saying that if the government that governs best is the one that governs least, then the absolutely best government is the one that governs not at all! Ridiculous as this is, there are intelligent people who seem to believe it. This brings us to the problem of an economic ideology fashioned by a distinguished department of economics at an equally distinguished Midwestern American university. There, a brilliant school of free-market economists provided a mathematically sophisticated theoretical doctrine that perfectly suited the wishes of those major American corporations seeking unregulated multinational corporate expansion. This is the Department of Economics at the University of Chicago, discussed in detail by Naomi Klein in *The Shock Doctrine: The Rise of Disaster Capitalism.*[2]

We can ask if it is possible to identify a false assumption that underlies the sophisticated models that have given support to those whose only concern seems to be the corporate bottom line. Many of us think this can be done. The critique is simple and unoriginal, but it seems to have been overlooked by some of the best-known economic modelers. This is the assumption that people can be trusted to be rational in their economic choices. To this, it is difficult not to reply, "Nonsense!"

Perhaps in Adam Smith's day, when producers and consumers alike were often local and might know one another, a more rational assessment by consumers could have been possible. Today, with generously funded, modern, psychologically astute marketing aiding huge multinational producers selling globally, can anyone really think that Adam Smith's model still applies? Indeed, further support for the nonrational factor in consumption comes from the role of advertising, in which the consumer's emotions are manipulated by many of the same techniques used in political propaganda, sometimes crafted by the same well-trained, well-paid people. "What should I buy?" is often a question that is answered for the consumer by the producer. How can an economist assume the consumer is rational when the best producers and advertisers know he is not? Yet much economic modeling assumes otherwise. "Shop 'til you drop" is a colloquial expression well-known to American consumers. The encouragement came from psychologically astute American marketers, though they alone cannot be blamed for how many consumers overspend wildly. Buyers also need to accept some responsibility too, and many do not.

It is worth noting that the same University of Chicago economists referred to above have won more Nobel Prizes in Economics than those in any other institution, academic or otherwise. While the problems discussed here suggest that economics may not yet have reached the pinnacle claimed for it as "queen of the social sciences," the ability and sophistication of these economists cannot be denied, and I hope most of us would accord them generally good intensions. A free-market counter position to the critique presented here could note that it is not from a professional economist, and therefore should be ignored. That is a fair criticism. To address it, we must ask if there is any rational academic study to support the claims that consumers and producers alike cannot be assumed to be rational agents.

Fortunately there is, and those who did the study have also shared a Nobel Prize in Economics, not by modeling but by devising and applying clever experiments in applied psychology. In *Thinking, Fast and Slow*, Daniel Kahneman discusses these experiments, in which student volunteers were given modest amounts of real money and many opportunities for investing and spending it.[3] These experiments revealed what any astute psychologist might have guessed. Many of the economic choices made by the students were not the optimum rational ones, whether their role was investor or consumer. The experiments were designed so that a purely rational economic choice could be established in all cases. This result suggests that human emotions play a large role in some of these decisions (which does not make then necessarily "wrong"). It means that human emotions can strongly affect producers and consumers alike, in ways that often do not optimize their purely economic choices. This work also suggests there may be serious limitations in current economic models, even the most elegant ones, if they are based on the assumption that producers and consumers are rational agents.

Moving from the personal level to the institutional one, I know no intelligent person who thinks the behavior of a stock market can be reliably predicted on a significant level of detail. Wall Street has developed a highly adaptive system for making swift adjustments to correct for private hunches and public passions. This still falls short of predictive power. To say this is based on some profound economic model that properly incorporates the irrational human factor that cannot yet be modeled is not credible. At least one former broker may understand the market's detailed unpredictability: Nassim Nicholas Taleb, author of *The Black Swan*.[4] Even economic modeler Paul Romer, a product of the same university that produced the current gold-standard free-market models, made no attempt to incorporate human irrationality into economic models after arguably improving existing ones by successfully incorporating the role of new, usually scientific, knowledge in a superior way.[5]

The sophisticated laissez-faire economic arguments of several distinguished economists at the University of Chicago, including several Nobel Prize winners, have served as a focus for this discussion of economics. The evidence collected by Naomi Klein suggests their application has proved generally harmful to the welfare of the great majority of people everywhere.[6] It would seem that, just as trickle-down economics and supply-side economics never quite led to the proverbial rise of all boats for others, then neither have the ideas of this "Chicago school of economics." If that is the case, one other factor has made regulation of these giant corporations difficult. To further their power, two American Supreme Courts one century apart have given these behemoths essentially the same protections under the Bill of Rights as are given to individual persons, which was discussed in chapter 8. Are some of us missing something here? Who are the real beneficiaries of all this activity? The ordinary citizen? It would seem not.

Application of these neoconservative/neoliberal economic theories, taken to the limit, resembles an elegant exercise in reducto ad absurdum, as noted above. We can ask why so many Americans in particular have accepted these theories. Beyond ignorance and media propaganda, their popularity has certainly gained greatly from the collapse of the former USSR, as already noted. Many intelligent liberals and conservatives alike agree the USSR collapsed largely from a bureaucratic debacle that was supported by a rigid authoritarian ideology. It was probably accelerated by the deliberate and dangerous, but effective, policies of the Reagan administration to break the Soviet's back economically, in part with an arms race the Soviets could not ignore and yet could not win. This policy was expensive and risky, requiring for success that the Soviets were civilized enough when cornered not to attack with nuclear weapons. Nevertheless, there is abundant evidence that this policy achieved its goal.[7] That said, and with a nod from a progressive to this well-documented conservative claim, the more simplistic conclusions some neoconservatives have drawn from the Soviet collapse could be ignored, if they have not also undercut support for needed government regulation and services and produced a regressive economy in the United States.

However, criticism of what may not have worked for the general welfare is easy compared to finding corrective measures that will. One example is achieving adequate government regulation, even when provided by law. Certain proposals for adequate regulation of freewheeling American corporations might work if it were politically possible to get them through Congress. Keeping good regulators in the government before they are lost through the well-known revolving door would help. When dedicated persons in regulatory positions know they can double their salaries by moving to the corporate side, small wonder that many of them do. An effective regulator must know what he regulates, so knowledge of the producer is needed. This argument is sometimes used to defend the revolving door. However, if a regulator's salary were doubled, he might stay put and become a more effective civil servant. While that is an attractive idea, gaining congressional support for it may be impossible, for the representatives of the corporate world are almost certain to vote it down.

Despite this current situation, the American government continues to employ a significant number of people in regulatory positions for modest salaries compared to what the private sector might offer them. These are people who are dedicated to their jobs and who are often effective in their performance. Just as impressive is the large number of people in the corporate world who do the productive work and maintain high standards of competence and personal integrity. Even at the top are CEOs who understand the need for structural as well as technical changes that they know will eventually come, one way or another. Until those structural changes occur, their hands are often tied as well. The point is simple. There may be no quick fix to the problem of reining in a mature capitalist economy with enormous inventiveness and productive capacity that also has a natural tendency to run wildly out of control . . . until there is a crash.

So if the prospects for significant progress toward solving the problems brought on by inadequately regulated multinational corporations are not currently very good right now, neither is the situation necessarily a global disaster . . . unless widespread ignorance and greed-driven ideology dominate the thinking of too many of our current decision makers for too long. Equally likely seems to be doing what the human race has done rather well for most of our history to date: surviving and muddling through, with a talented and/or lucky few doing very well and the majority mainly getting by but still finding life worth the effort. In principle, things could be better, but they could also be worse. The long-recognized but still-unsolved structural problem of mature corporate capitalism will be difficult to solve, but that increasing public pressure to advance toward more socioeconomic justice will continue is certain.

To the problems caused by socioeconomic injustice there is the further challenge offered to progressive modernization posed by reactionary religion. This category excludes the tolerant religious who favor a "live and let live" approach to life, who often support complete institutional separation of church and state, and who are as strongly opposed to a theocratic society as any nontheist. Nevertheless, there remains a large number of the religious who are not tolerant on these issues, and they have recently become politically more active in many parts of the world. When the intolerant religious reactionaries combine their political influence with those favoring unregulated free-market capitalism, as seems to have occurred in certain American political circles today so that both might gain power, it is hard not to see a perfect storm in the making. If military force should be used to further the goals of both, the cyclone will only gain further strength and destructive power.

Thus we begin to appreciate the magnitude of the contemporary socio-

economic problem, where power-oriented political advisers have orches-
trated a marriage of convenience between the economically well endowed
and the spiritually insecure to move politics in a very troubling direction.
While this arrangement claims to satisfy people's needs for both bread and
divine favor, it does not negate the insights of author Chris Hedges, who
notes the similarities with traditional forms of fascism.[8] There, freewheeling
corporate power, authoritarian government, militarism, and quasi theocracy can put an effective clamp on a liberal society, even one that maintains
some semblance of democratic elections.

Yet these developments reveal only one side of the contemporary
picture. The other side concerns the role of the people in an effective
democracy. To blame everything on corporate moguls, clever marketers and
propagandists, autocratically inclined executives, militarists, and religious
reactionaries overlooks the role the people can still play. Should any democratic society degenerate into a militant fascist theocracy, the people are not
innocent and must share the blame. I would not predict such an outcome
of current trends that still bear watching, but the subject will be revisited
in the final chapter. Perhaps the worst attitude any American citizen could
assume now, outside of becoming depressed and doing nothing, is that
all is well and complete recovery is just around the corner, if we will only
return to business as usual.

THE INEVITABLE FAILURE OF REACTION

Who are the reactionaries? I have used the term frequently in previous
chapters but have never attempted to define it, since, like many words used
today to describe political and/or religious positions, *reactionary* means
different things to different people. So my hope was that the meaning
would be clear from the context in which the term was used. However,
whenever the word *reactionary* was used, one characteristic of a political
and/or religious reactionary was always present. That characteristic is a
tendency to look to the past for inspiration and guidance.

Looking backward is not always a bad thing. The first chapter in this book briefly reviews the history of humanism, which to be understood must be viewed in the broader context of the times in which humanistic attitudes and values developed. Philosopher George Santayana is often quoted as saying that those who cannot remember the past are condemned to repeat it, to which history gives some credence. On the National Archives building in Washington, DC, appear the words "Study the Past" and "What Is Past Is Prologue." Isaiah Berlin, introduced earlier in this book and famous for his comprehensive knowledge of history, frequently comments on the dangers of *any* rigid ideological position, including some we might ascribe to "doctrinaire liberalism."[9] I generally agree with these viewpoints, except for the implication that the past is *always* prologue. That position is not supported by developments since the eighteenth century, and I think it will be at least partially negated by future developments in the biological and neurological sciences. Chapter 12 comments more on these future developments.

It is important to note one other almost-universal characteristic of reactionary individuals. Reactionaries almost invariably exhibit a rigidity of mind that is difficult to penetrate and that is often associated with a passionate adherence to an ancient, typically religious, text that is claimed to supersede all other sources of guidance for human activities. This characteristic not only tends to instinctively block most progressive ideas; it also makes many reactionary positions dangerous in the contemporary world. When an opposing position is presented to the reactionary believer, a common reaction is not only to deny it—even when the position is based on reliable evidence—but to support violence as a means of sustaining a traditional outlook. The examples should be familiar to every person following current events in the world today.

Most conservatives are not reactionaries, and reactionary views should not be automatically assigned to them. Few political scientists in a democracy would argue that a healthy society should adhere to a doctrinaire kind of liberalism that attempts to suppress conservative dissent. Intelligent conservatives give society a needed check on the possible negative consequences of some of the ideas enthusiastically proposed by liberals of limited practical experience.

One example of such an idea is an early public-housing program that followed World War II to improve conditions for people living in inner-city slums. The idea then popular was that giving people a more attractive environment would change their lives, so the slums should be razed and high-rise public-housing projects built in their place. Some of these projects were such disasters that *they* had to be razed, like a large housing project in St. Louis—my former hometown—that corrupted a once poor but functional neighborhood.[10] We know why this idea failed, since we now know that the human mind learns by changing physically over time, not instantaneously (see chapter 5). In the case of these ill-considered public-housing projects, the environment did not change the people. The people changed the environment. Overcrowding of generally poor and ignorant people often proved even worse than a distributed slum by providing more opportunities for criminal activity. We need sober-minded conservative critiques of such ideas before launching *major* public projects. Small trial projects are another matter, and the lesson in this case seems to have been learned.

Another, perhaps deeper, reason the conservative mind is an essential part of a healthy society today is the tendency for some progressives to ignore our still-elusive but stubbornly real human nature. Biologically evolved human nature has often proved to be the bane of the radical leftist ideologue. Natural human propensities frequently interfere with theoretical schemes for the rapid perfection of human society. The "new communist man" proved to be disappointingly human to many cultural idealists. However, that, too, is now generally recognized, arguably even by many of those who may still call themselves communists or at least Marxists.

In contrast to this valuable conservative skepticism, reactionary attitudes and movements tend to exhibit a passion-driven intensity that can give people strength in opposing all progressive developments. This intensity can also make them dangerous if they exhibit a penchant for violence. The highly regressive Nazi movement discussed in chapter 8 is only one example. If reactionary movements are common in the world today, why can it be said with confidence that reactionaries will eventually fail? The argument for this statement is entirely pragmatic. It also requires taking a global view and looking ahead, beyond the time of most of us living today.

The first part of the demonstration involves the strong will to live and to thrive that is built into our genes and that is a product of our evolutionary development. Most people are biologically inclined to seek a decent secular life for themselves, even if, like religious extremists, they have been taught otherwise. Suicide bombing and other forms of self-immolation are not new to history, but there is both contemporary biological as well as historical evidence that the practice is not and never has been popular, except under highly abnormal conditions of extreme real or perceived adversity. Courage in accepting mortal risk in service of pursuing a demonstrably ethical goal is generally regarded as a mark of a healthy mind and the highest form of character. That is hardly true of one who finds life onerous and consciously wishes to depart it. In sum, what Richard Dawkins calls "the selfish gene" is working against suicidal tendencies in people. It is not weakness but a strong and healthy survival instinct that induces us to jump back quickly, without thought, when a high-speed vehicle is bearing down on us. When the antihumanistic or destructive sides of reactionary attitudes and movements become apparent, a growing number of people will *instinctively* turn against them.

The second part of the demonstration concerns the nature and the value of science. This book has already summarized the argument in several other contexts. Science gives us the most reliable knowledge possible of everything in the natural world, which is sometimes called "the natural order." If there is no other (supernatural) order, then science can give us the most reliable knowledge we can hope to gain of *everything*, including the working of our own minds. That states *the nature of science*. The *value of science* is even easier to summarize. *Science works.* Science provides exactly what is claimed for it, which, as the above statements show, is reliable, comprehensive knowledge. That completes the demonstration of why reactionaries are fighting a losing battle.

On the basis of this argument, reactionaries are bound to eventually fail, because they are pursuing dogmas and doctrines based on tradition and emotion, not on reliable knowledge. Reliable knowledge combined with the human will to live and to thrive is the surest way to a better future for people everywhere. These offer the best guide to ensuring both survival and a more satisfying world for our descendants. That is exactly what the Enlightenment was all about.

PROGRESS UNDER RESTRAINT

To reassess the primary goals of the historical Western Enlightenment, it is useful to restate them. Introduced in chapter 1, they emerged in the seventeenth and eighteenth centuries as a grand synthesis of humanistic attitudes and values that began developing in earnest during the Renaissance but that had antecedents in the classical world. The primary goal was the development of a better secular world for humankind as a whole. This was in contrast to a world organized primarily for the convenience of a ruling minority, or to satisfy the wishes of a postulated God. From a secular perspective, the primary goal was profoundly ethical, humanistic, and universal.

There has been strong opposition to this secular orientation from the start, especially among those described in this book as the political and religious reactionaries, who tend to resist any change that threatens the status quo. This opposition is especially strong when prevailing economic interests or traditional religious dogmas are threatened. A strong anticlerical movement became associated with the historical Enlightenment, even though not all of those supporting the goals were atheists or even deists. While the primary goal was inspired by humanistic ethics, the means to reach it were identified as the pursuit and application of science and reason.

Chapter 2 and following chapters reviewed how well, or how poorly, the goal has been achieved. The picture is mixed. One thing that stands out is how optimistic and naive many, though not all, of the Enlightenment thinkers were when evaluated in light of what has been learned since. A notable exception was Voltaire, who is often criticized today for not being radical enough; but a convincing case can be made that he was both shrewder and more realistic than, for example, Condorcet or the group that gathered around Baron D'Holbach's salon.[11] Politics has always been the art of the possible. Voltaire understood patience and compromise, but he also had a remarkable sense of timing in knowing when to move aggressively and effectively for progressive change.

Though the more visionary goals of the Enlightenment were not to be realized easily or quickly, much has been achieved since the eighteenth century. Without the Enlightenment, the progressive American Revolution probably would never have occurred. Democracy has struggled ever since, here as elsewhere, but even most American political reactionaries and their counterparts in many other countries now accept the ballot as the legitimate way to gain political power. The recent eruption of the Arab Spring illustrates how this idea has gained worldwide appeal. Even the economically successful contemporary Chinese mandarins who call themselves communists and the current Russian czar cannot ignore the appeal of democracy today.

There is also no denying the success of corporate capitalism in producing a staggering variety of goods that have lifted hundreds of millions of people to a higher physical standard of living. If that economic system were generating no ancillary problems, there would be little reason to criticize it, but even in light of those problems, the success of capitalism as an engine of production cannot be denied, as the aforementioned Chinese leaders recently discovered.

Perhaps the most impressive achievement of the Enlightenment for thinking people has been the dramatic rise of modern science and the change in their perception of the world that has resulted. This change has definitely been in the direction of developing a more naturalistic view of the universe and a more secular view of human society. This success has stimulated a strong reaction against certain conclusions of modern science, as many people have awakened to the impact of these science-driven ideas on their more traditional views of the world. This is a tribute to the prestige of science, even among those who do not always like it.

Despite this reaction, there is general support for most scientific research, thanks to an almost-worshipful attitude toward modern technology and especially toward modern medicine. There is a general, if not universal recognition of the importance of science in providing the knowledge for continued improvements in these areas. The easiest example to grasp is the research in the life sciences that ultimately leads to better medicines that cure or mitigate the effects of human diseases. Today, this research is based on the evolutionary model first articulated in detail by Darwin. Religious reactionaries may try to deny this science, but their efforts are certain to prove futile as more of them realize why it is both valid and important.

It might seem on the basis of these remarks that most goals of the original Western Enlightenment have been largely realized, and that all that is needed is "more of the same." There are those who would support that view, especially in the United States. However, this extreme optimism, often based on a near worship of technology, could be dangerous. The following examples are drawn mainly from the American experience as a global power, but the concerns arise from the historical experience of many nations.

Since the American Revolution, a clear victory for Enlightenment views, the relative status of national wealth, power, and population that existed in America during the eighteenth century compared to the major European countries has been reversed. Today, the United States has assumed the role of a major world power and the attendant responsibilities that go with that position. This role reversal invariably has had consequences. Not everything can remain the same when a society that was once concerned primarily with its own internal development has become involved with every other nation on the globe. What those accommodations should be is beyond the purview of this book, but some of the consequences must be noted, because they illustrate how challenging it is for a nation that owed its birth to the Enlightenment to continue pursuing Enlightenment goals.

The first consequence concerns the Enlightenment goal of separating the *direct* participation of religious institutions in government. Regardless of an American's personal views on religion, few educated people think the country is on the verge of creating an entirely secular society today. Most of us who are not religious would be satisfied if the *institutions* of church and state were separated, which does not mean that religious people would be prevented from expression of their views in the public square. Yet even this seemingly modest proposal for institutional separation has never been realized in the United States and has been strongly opposed by at least one current Supreme Court justice. It would seem that this product of the Enlightenment, America, has grown up to be a rather cautious conventional adult in this area.

Second, we have already noted the problems of juggernaut technology that is resistant to course corrections away from doing what its directors already know how to do, often opposing effective regulation in the guise of an anachronistic free-market ideology dressed up in fancy mathematics. These free-market arguments ignore the more convincing critique of unregulated mature capitalism and the historical evidence that it naturally tends toward monopoly, economic imperialism, and war. Creating and maintaining a just and sustainable global economy today is a problem that remains unsolved.

The third concern is the political consequence of a global power's need to maintain a powerful standing military force. Fortunately, relative isolation by land has given America the advantage of emphasizing first sea power and more recently aeronaval power extended into near space. From the histories of Athens to England, compared to Sparta and Spain/France/Germany, respectively, historians have noted that nations maintaining large standing armies in their own territories are more vulnerable to tyrannical governments than those not so compelled. Nevertheless, the American military establishment has become an enduring colossus. Historical circumstances since 1939 have required a powerful American military force, but it is useful to remember that there are historical dangers of having one. The military mind, with notable exceptions, tends to be socially conservative and hierarchical. That may be changing today, but it does so always within limits circumscribed by the requirement that a military force must be effective in combat. Of greatest concern, a powerful military is like a big hammer. Depending on who is in charge, the temptation to reach for that hammer when a problem arises abroad may increase in direct proportion to its size and availability. The consequences of this for maintaining a liberal democracy could eventually become a problem.

Fourth, the very success of science has energized efforts to politicize it, and in some cases even to distort public perceptions of it. We can hope that certain current efforts to impede scientific research will be temporary and soon overturned, when the results of science conflict with contemporary religious dogmas or economic interests. Limitations on allowed research procedures with stem cells have caused delays among researchers, and progress in several medical fields has been slowed. Greater public awareness of this situation would probably accelerate progress here. The greater danger in this case and others is the question of who is best qualified to evaluate the results of scientific research: the scientists who do the research or nonscientists with an ideological position to defend. Giving the authority to the latter agent compromises science itself, subordinating it to politics. Yet the American government has recently done just that, in the areas of stem-cell research and climate science.

Many other examples of societal failures to reach a more just, stable, and sustainable world have been given in previous chapters. Referring again to Jared Diamond's book *Collapse*, the author's list of ten major global problem areas is sobering.[12] Most of them concern severe stress of the planetary environment. One problem cited is barely controlled population growth. Most of the others involve resource exhaustion. Diamond attempts to present a brave front and does not identify any one item on the list as the most dangerous. Yet continuing population growth fed by uncontrolled growth in production stands out as driving all the rest. It requires a great leap of faith in techno-optimism to assume that all of this can be brought into a healthy balance if all large current technologies are permitted to proceed and grow with no effective regulation.

The conclusion of weighing the benefits and the drawbacks of the developments of the past two centuries gives us a mixed picture. It is time to offer an assessment of how well Enlightenment goals have been realized today, and to ask if the goals themselves may need some change. Is it possible that the more optimistic thinkers of the Enlightenment significantly overestimated the human potential for creating the kind of just, prosperous, and stable world they envisioned? That is the question the final section of this chapter addresses.

THE ENLIGHTENMENT TODAY

Many reasons for the failure of people to achieve a still more civilized state have already been discussed here and in previous chapters, so repetition is unnecessary. However, that leaves unanswered the question of human capacity to realize the more humanistic kind of secular civilization envisioned by many prominent eighteenth-century thinkers. A large number of intelligent people throughout history have offered a negative answer about human capabilities to produce such a just secular society, and have offered evidence and arguments to back it.

Traditional Western Christianity remains skeptical about overall human improvement through human effort alone, and this is often supported by impressive scholarship. Though not all Christian denominations invoke the bleak doctrine of original sin in trying to explain human failures, most of them claim that without help from a "spiritual" source beyond human capabilities, the lot of humankind cannot be greatly improved. Outside of the West, the godless doctrines of traditional intellectual Buddhism—in contrast to popular varieties more open to comforting gods—are hardly a celebration of secular life. The "four Noble truths" emphasize all that is painful about life and the desire for its satisfactions. Most people would agree that religion is such a powerful part of human culture because life is often difficult and is not infrequently hard for many people to bear.

The counter position of the Enlightenment argues that by overcoming ignorance and superstitious fears, and by applying that knowledge to life-enhancing skills, secular life for more people can become inherently fulfilling, and the human race can move forward. That is the position of secular humanists today, and there is abundant evidence to support it as well. How can the contradiction between these pessimistic and optimistic views be resolved?

We have noted the close association of the optimistic eighteenth-century Enlightenment vision for a higher civilization with science. Allowing for an excess of optimism among some of these thinkers, their idea that science holds the key to reaching the goal remains sound. The argument already given here is strikingly simple. If the goal is to understand the natural world of which we are part, science works and superstition does not. If there is no world outside of the natural world, science can eventually provide the knowledge we need, including the generation of supernatural ideas that will then be seen as fantasies of the human mind/ brain. The struggle to reach these higher levels of understanding may take much longer than the original proponents of this approach realized, but that does not mean that these problems will never be solved. In principle they can be solved, and even in practice some remarkable first steps are now being taken. That it will take ability, patience, and courage to reach this understanding goes without saying, but those are qualities that many human beings have already exhibited throughout recorded history.

One more observation is needed to complete the picture. The assumption is that the human race can and will produce the people to carry this project forward without becoming unduly discouraged and regress to conventional ways of thinking and acting. That assumption must be examined more carefully. First, it is not necessary that everyone or even most of the people must answer the call to provide the needed new knowledge or to transmit it to a larger public. For society to function at all, there will remain many other vital roles to fill, as there are in our already-complex society today. Only a small fraction of a population needs to be involved in the creation of new scientific knowledge, just as only a small fraction of a population will have the talent and/or the desire to create the arts without which our societies would be dull and colorless.

Even more convincing is that the human race is the product of three and one half billion years of biological evolution on the planet earth. That life has continued here for this long and has developed life-forms as complex as we are is a tribute to the toughness of life. This has produced in the human being an innate intelligence that, properly applied, should greatly increase the odds for avoiding the extinction that has eliminated a large fraction of other, less adaptive former life-forms, most of which succumbed to overspecialization. The genetic makeup of the human being has given us a strong will to live, to explore, and to thrive. Many of us have succumbed to ill-conceived ventures before, and more will no doubt succumb in the future, but as the progress toward a better life for a larger fraction of the human population progresses, aided and abetted by scientific advances, this will to live, to explore, and to seek that better life is likely to be irresistible.

The most probable and likely outcome of this process is exactly what the philosophers of the eighteenth century had in mind: a higher state of civilization for people over the globe that would provide a more fulfilling secular life for a much larger fraction of the earth's population. The vision was correct, even if early estimations for the time required to realize it were seriously flawed. Today we understand why their estimate of the time required was wrong. Many of these highly rational thinkers seriously underestimated the role of human emotions as co-determinants in human activity. (A few, like David Hume, did not make this mistake.) This excessive trust in reason *alone* has now been shown to be an error. Far fewer educated people today are likely to make this mistake, though one can still find in academic circles some who are puzzled as to why there are still intelligent educated people who believe in things for which there is no evidence.

Thus, the time frame for realizing the kind of society envisioned by Enlightenment thinkers is surely much longer than the more optimistic of them hoped it would be. Also, their faith in the power of reason to quickly convert people to only those views that accord with evidence and reason was excessive. This is not altogether good news, for the formidable inertial forces of reaction still arrayed against the desired progress are unlikely to be overcome quickly or easily. A good deal of human suffering remains likely in the process, though one can always hope for a better outcome.

However, for those willing to take comfort from the longer view that extends well beyond the lives of those living today, we can say that the vision of a more enlightened society in which a much larger fraction of the human race will enjoy more fulfilling secular lives is not only capable of realization but very likely to be achieved. *The grand ethical Enlightenment vision was valid and should be retained.* It is arguably the most noble program humanity has yet devised for a higher state of human civilization. *Even the means proposed to achieve the goal are valid and workable,* as long as we take into account those stubborn human emotions that often block progress but that sometimes also promote our very survival. *We must remember to be patient and to accept that no road worth traveling is likely to be easy, even for dedicated problem solvers.*

That, in sum, is my assessment of the Enlightenment.

I leave the last word to a historical figure too pragmatic to be a pure idealist, too smart to build a grand universal philosophical system, and too ordinary and self-promoting to be heroic until it really counted. Yet Voltaire may have done more than any other person in history to reduce the influence of the world's then most powerful religious institution, which both educated him and often courted him until the end of his life. When concluding his best-known story, *Candide,* Voltaire's protagonist of multiple misfortunes replies to the silly philosopher who insists that this is still the best of all possible worlds. "That's well said." Candide replies to Pangloss, "but we must cultivate our garden."[13] There is still plenty of work to do.

CHAPTER 12

TOWARD A BETTER FUTURE

SUMMARY OF *THIS* ENLIGHTENMENT THESIS

This book has noted from the start that the thinkers normally associated with the eighteenth-century Enlightenment were primarily motivated by a humanistic concern for the welfare of people everywhere. As the means to that end, science and reason were recommended. Thus, the movement was profoundly ethical from its inception; secular in its means; global in its intended outreach; and in view of the often-repressive power of organized religion, strongly anticlerical.

Because major societal progress was arguably nonexistent in the medieval world of the West, and hardly dominant in the philosophies of East and South Asia, *the idea of progress* was central to the Enlightenment. This has persisted to this day, and the terms *progressive* and *reactionary* have been used liberally in this book to distinguish between those whose mental orientation is predominantly toward the future or instead toward the past.

It has also been noted here how optimistic many of the early Enlightenment thinkers were. Many thought they had seen light at the end of a rather long, dark tunnel. The American Revolution gave impetus to the view that humanity was on the verge of endless progress. Unfortunately, the immediate consequences of the violent French Revolution suggested that a more sober view should be taken. In the assessment of many scholars, the optimistic original vision has been further modified by several major events of the last two centuries. This cautious thinking continues today, though there remains plenty of politically motivated "cheerleading" to sustain the original optimism. Much of this proves to be faux optimism in the service of political and other agendas, but this is a normal part of a healthy democratic process.

The question arises as to why the original optimism has been greatly tempered; in other words, what are the underlying causes? Ignorance, superstition, and the inability of the human race to tame the currently most successful engine of production—capitalism—to achieve socioeconomic justice and global stability have been identified as major causal factors. There are others, and some of these have also been discussed in several of the previous chapters. However, as long as the above three causal factors persist in their present form or, worse, regress to an earlier stage, it is hard to imagine many of these other problems being solved. The associated challenges include the need for a clean environment, a healthy ecosystem, population stabilization, improved human rights, and a world government strong enough to prevent major wars resulting from resource exhaustion or environmental collapse. Thoughtful people recognize how much all these challenges are interconnected.

Reflecting on this situation, I have come to the conclusion that no one causal factor is as serious as the crippling ignorance of a large fraction of the world's population, including the population within the current United States. Superstition is the child of ignorance, and the structural problem of mature corporate capitalism could be solved without destroying capitalism's admirable entrepreneurial spirit or its impressive productivity . . . if and only if enough people understood the problem. We have noted that ignorance of two kinds can be easily identified: ignorance of what is already reliably known, and the ignorance we all share of what is not yet known. Thus, both education and further research are needed, the one to teach critical thinking and to effectively transmit the knowledge already available, and the other to gain further essential knowledge where it is still lacking. We have also noted the passionate opposition of many political and religious reactionaries to achieving these educational goals.

It would seem from the above remarks that the current state of humanity as a whole lies somewhere within a range from seriously challenged to marginally desperate. Indeed, if no major changes occur in the next half century, with some beginning soon, I would confidently make that assertion. That prompts the question of what might help us avoid another disaster of global proportions in the current century, one at least as devastating as the two world wars of the previous one. We can add that the next major global disaster seems more likely to be environmental than military, though military response then becomes one possible outcome, an eventuality that American military intellectuals are already taking seriously.

On a more hopeful note, this book notes a number of developments that increase the odds for progress but invariably comes down to two in particular. One is a mitigating action that will require a major deliberate effort soon, in a time frame sufficiently short to prevent tragedy in the near term. The other is a genetic attribute of people that functions automatically, though its operation is so silent that many people do not realize the force for progress that it provides.

The first of these concerns a determined political effort to increase effective education and to continue supporting further research. We have noted in several chapters the value of communicating reliable knowledge to overcome the problem of crippling ignorance. Chapter 4 demonstrates why science provides the most reliable knowledge we can gain of everything in the natural world and the best way to obtain more. It is impossible to overestimate the value of this available knowledge, while acknowledging the need for further investigations, especially of ourselves—the main actors in the human drama.

We already possess considerable knowledge that could be usefully applied, including a collected wisdom on human behavior and vulnerabilities that is often transmitted by religion when it remains humanistic and tolerant and ceases to emphasize ugly traditional notions like original sin. (Evolutionary biology, once understood, provides a much more humanistic, and scientific, explanation of human vulnerabilities.) Unfortunately, there is no guarantee that enough of the world's population in any major society today will quickly attain both the necessary knowledge and the willingness to apply it to prevent many destructive outcomes in the near future. The opposition to reaching even the goal of effective education comes from passionate individuals who remain determined to turn society back to a darker age, based on a literal reading of texts written centuries ago and already discredited by science and modern scholarship.

Despite these constraints on progress imposed by proponents of reactionary views, there is a second mitigating factor that gives us hope. This is a nonrational (not necessarily irrational) natural human faculty, which is part of our genetic heritage. We have already noted that what Richard Dawkins calls, in popular language, "the selfish gene" gives each human being a strong preference for life and personal fulfillment. Each of us has nothing to do with this extremely life-enhancing biological propensity. It is built into our genes and is a product of over three billion years of biological evolution. Among the other faculties our evolved human nature has given us, however accidentally, are a high innate capacity for intelligence and a still only partially understood (on scientific grounds) innate capacity for empathy. These faculties, properly developed, are the basis for intelligent action and good character, without which civilization would be impossible.

When the immense value of the reliable knowledge provided by science and the innate desire all people have for more fulfilling lives are combined, the probability of human survival and progress becomes much greater. The recent Arab Spring seems to support this view. These two mitigating factors remain the likely basis for why most people continue to find life worth living and devise ways to "muddle through" during even the worst of times. This was noted in the earlier discussion of the Black Death that struck Europe in the fourteenth century. We can reasonably hope to do much better today, even with the constraints imposed by reactionary thinking, and we have a strong potential for doing so. Our innate biological attributes are one reason many people think there is little likelihood for a *major* exchange of nuclear weapons. This view is supported by the responsible actions of a repressive Soviet government when it was faced with a decision to attack or capitulate during the arms race of the 1980s. Rogue nations under crazy dictators are another matter, but the world seems to be watching them carefully.

The conclusion reached in this study is simple and, I think, persuasive. The humanistic goals of the Enlightenment remain valid and should be retained and reemphasized today. The proposed means to achieve the goals are also valid. The application of science and reason, not superstition and/ or ideology, are the best means to ensure a better future for people everywhere. To that must be added an honest acknowledgment that progress toward reaching those goals has been painfully slow, compared to the hopes of progressive thinkers living at the time of the American Revolution. The inertial forces of ignorance and reaction remain strong, and those adhering to anachronistic positions today will not alter their views quickly.

Nevertheless, even with these people, there is reason to think that their descendants will one day think much differently and support the more progressive outlook. There is an encouraging American example of how people can change and embrace progressive change. Some of the New England abolitionists of the nineteenth century descended from progenitors who profited from the slave trade in an earlier era. What better example could be given of the importance of education and a firm grounding in ethics, along with the courage to act accordingly?

To that, one further observation is in order. The ethical humanistic nature of the Enlightenment has been noted frequently here. Ethics as a subfield of philosophy is studied by only a small fraction of university students today. That is unfortunate and may reflect the current focus on the immediately practical and the technical in most contemporary education, especially when employment issues loom large for graduating students. To give one example where a better understanding of ethics would be valuable, I have known intelligent people so enamored of what science has *already* revealed that they will dispute the validity of the previously mentioned naturalistic fallacy (see chapter 4). Recalling that the naturalistic fallacy states that we cannot derive a prescriptive system from a descriptive one, that is, get our ethics from science alone, this is certainly still true. A future convergence of scientific knowledge in all fields may someday bring us close to reducing the relevance of the naturalistic fallacy, in the spirit of E. O. Wilson's intriguing arguments in his book *Consilience*.[1] However, whether we will ever reach that level of detailed understanding remains an open question, and to suggest we are close to achieving it today is demonstrably absurd. Today, we hardly even understand ourselves on detailed scientific grounds.

Interestingly, I know no working scientists who would claim that science has replaced ethics today, and neither does Wilson. At least for today and the foreseeable future, the field of ethics still lies at the foundation of prescriptions for personal action and for law. A better understanding of ethics would make this clear to more thinking people. The final section of this chapter will give an example of where a neglect of ethics has influenced the failure of the United States to initiate serious action to reduce global warming.

LOOKING FAR AHEAD

This section is imaginative and speculative. Nevertheless, reflections on distant future possibilities can be more than an enjoyable parlor game. One thing we can say about the future with confidence is that it will not be identical to the present or the past. From Heraclitus among the Greeks through Darwin in the nineteenth century to the expanding universe of today, change is a fundamental property of reality on all levels excepting possibly only the most abstract. Perhaps fear of the uncertainties associated with change is the deepest anxiety of the reactionaries, but that change will occur is certain. So this section speculates on some possible changes, not in detail—which is not the spirit in which this book is written—but in some interesting general areas. Much of what follows here will concern the *distant* future, and I regret that some of it may offend those who think the human future should resemble human life today.

To begin this discussion, let us assume that the current major problems considered in this book have been solved and the major goals of the Enlightenment have been met. The human world has advanced to a higher stage of civilization. There is effective world government to preserve peace among the nations, changing climate is manageable, and socioeconomic justice has been achieved along with universal human rights, which in turn have promoted a stable population fostered by universal equal rights for women. The authority of reliable knowledge reigns supreme, and authoritarianism is dead, especially the more extreme forms associated with brutal father gods and their more violent supporters in the current human community. The environment is productive and attractive, and the ecosystem is healthy. Most thinking people would like to live in such a world.

Along with developments that might lead to this future state of civilization, there are sure to be changes in the human genome. As toxic industrial chemicals have already entered our environment, subtle changes in our genome are probably occurring today. Biological evolution does not stop with the human species, and these slowly accelerating changes may have begun with the rise of modern science and the Industrial Revolution, even if they were imperceptible. While many astute thinkers have concluded we are still far too ignorant to engage in major "genetic engineering" of the human species, there is little doubt that day will come.

By the time the kind of enlightened society described above arrives, genetic manipulations beyond those to prevent inherited diseases will probably be under way. Just as *Homo sapiens* gradually replaced the Neanderthal and the earlier Cro-Magnon species, while possibly interbreeding with them, the day may eventually come when future biologists declare that our kind has evolved sufficiently to justify a new official description for our descendants—*but they will still be our descendants*. Who could object to our distant progeny being not only more intelligent than we are but also much more empathetic and compassionate? This latter development could largely address Freud's insightful concern expressed in *Civilization and Its Discontents*, in which the author notes presciently that the biological human being does not live altogether comfortably in the civilization we have created, even though the benefits of civilization to the majority clearly outweigh the drawbacks.[2] Continuing to evolve beyond a somewhat-natural barbarism has to be a good thing.

Having posited this more advanced civilization, we can ask, "What might lie beyond it in our descendants' future?" Of course, any answer becomes highly speculative but has the value of expanding our imagination into a universe that we have only recently discovered is staggeringly vast and complex, and for which theoretical physics and cosmology have yet to pin down that elusive "theory of everything." Nevertheless, as Einstein expanded on Newton and as Heisenberg expanded on Einstein, the achievements of all three currently still stand within the range of their assumptions, properly stated. If still incomplete, this science has created an interlocking web of self-consistent relationships that has, along with corresponding observations, given us views of the natural order far more astonishing than any model imagined by the ancients or attributed to any primitive gods. On this basis, certain future conjectures can be made.

We can begin by noting that, thanks to science, much can be done when the need arises to meet future major natural challenges to our collective survival. A good example is how our descendants will likely prevent the otherwise-inevitable impact of a future killer asteroid, one large enough to set back the entire global civilization. The database on asteroids and comets is already extensive and growing, and the technology may soon be available to divert the orbit of a large near-earth-crossing asteroid if it is detected early enough. There should be no future Chicxulubs and no corresponding large-scale species extinctions due to asteroid impact.[3]

Some people might still grieve over the impending end of continent-building plate tectonics, estimated by geophysicists to occur in one hundred million years or more. Eventually, the driving planetary heat engine powered by radioactive decay will die out. Erosion will eventually produce a global ocean, and we might need to live in or at least with a newly created water world. However, if our descendants are not smart enough to adapt to that situation by then, Charles Darwin, were he still with us, would probably be ashamed of them. For the gold standard of worries, there is the inevitable death of our nurturing sun, and with its expanding atmosphere, a baked earth . . . in approximately four billion years. Will our descendants still be stuck on earth at that time?

The point is not to trivialize major terrestrial or cosmic hazards, even extremely remote ones. Some of them are real and potentially catastrophic. Instead, we note that our knowledge of them comes from science. It is that same science and the technologies based on science that have given us the means of dealing with those hazards that are most likely to affect our descendants in the not-too-distant future.

Among the more down-to-earth issues already starting to confront us is that of how humans will link with increasingly intelligent robotic machines that may become sentient. I will skip the question of how human neural networks and those created by humans can be made compatible, and whether or not sentience, once it is better understood in humans, can be induced in intelligent machines. (Or whether the intelligent machines can generate it on their own!)

Let us assume that the ability to reproduce and to feel pleasure and pain will eventually become attributes of intelligent machines that our descendants will build in the future, and consider the possible consequences for both. We can start by asking what kind of linkage between humans and intelligent machines we would like.

The obvious answer is that we want our super-machines to serve our wishes and not be hostile toward us. This does not justify making slaves of these machines if they are sentient, any more than we are justified in cruel treatment of nonhuman animals. Thus, on ethical grounds, we can already identify two major constraints on how we should develop these machines. First, they should be developed in parallel with human development, to ensure mutual compatibility. Second, they should be granted freedom of action compatible with their own development. Satisfying these two constraints simultaneously may prove challenging, but it remains a problem that lends itself to logical mathematical treatment. Since we are the creators, attention to these questions will become increasingly important as intelligent machines develop further. Only if this is not done could some of the science-fiction horror scenarios become a serious problem.

People intelligent enough to create these machines are also clever enough to build in the necessary constraints. If this seems incredible, it is not fundamentally different from the ethical questions that our descendants will face when engaging in future genetic modification of the human genome to enhance our descendants' innate intelligence and empathy, and probably other attributes as well. Most of these problems lie in the future but require consideration now to prepare those who follow us and will perform this work. Some of these questions are arising today. One is the question of using "smart pills" to enhance intelligence, especially if their only effect is to make our querulous species smarter. Smart brains alone do not make us human.

I will end this section with a few speculations on how our descendants might seed the local group of planetary systems with *their* genetic material and initiate a long-term process for transporting advanced life beyond the earth in some future epoch of human migration. This can be viewed as reminiscent of the "out of Africa" exodus that began the human settlement of planet earth. I assume that two developments already under way have by then advanced to the following advanced stages: (1) the human genome has changed dramatically, and our descendants are both far more intelligent and far more empathetic than we are today; (2) humans have developed and have learned to live harmoniously with highly intelligent, empathetic, sentient machines, often called "intelligent robots" today.

The problems our descendants must solve to seed the galaxy are so difficult that it is only possible to mention some of them now, noting that they defy solution today and may prove much more difficult to solve than we can currently imagine. Nevertheless, if a person had asked Greek philosophers of 400 BCE if humans would one day walk on the moon, many might have answered in the negative—and that was only ~2400 years ago. The cave artist(s) who did the remarkable thirty-thousand-year-old paintings of Chauvet, France, would likely consider the questioner insane.[4] This spectacular emigration scenario is hardly in our near future, but to deny it to our descendants of the very distant future would seem unnecessarily restrictive and pessimistic.

The local stellar group that could be a candidate for such a venture is often defined as lying within thirty light-years of our solar system. If a suitable planet were located at the edge of this group, and it became possible to send well-constructed vehicles at an average speed of one half the speed of light, the vehicle could complete the one-way trip in 60 terrestrial years, and a round trip in 120 years. While achieving such a velocity is currently impossible in practice, it violates no known physical laws. There remain major problems of effectively communicating conditions on the planet over multiple current human generations, and of how such a planet might be made a suitable home for future humans. The effects of relativity at such speeds alone pose a major problem. Yet doing this is not ruled out by the fact that many long human life spans might be required to implement each step, from reconnoitering the planet to preparing it for settlement. Once the formidable physical problems have been solved, solution of the problem of sustaining future human life during a long immigration flight could prove surprisingly simple by comparison. To see this, recall that we are assuming that our descendants will have greatly exceeded our faculties in two critical areas and that our robotic "friends" are available to travel with us on the immigration flights.

The scenario is straightforward. Live humans would not need to make the journey. Only the eggs and the sperm cells of the future immigrants would need to be sent. Properly maintained under controlled conditions, brought together at the appropriate time for conception by robotic techniques, birthed and nurtured in a similar way by properly trained (and suitably patient) robotic nurses to young adulthood, the planetary pioneers would be at an ideal age for their great adventure upon landing.

It is probably best to leave this imaginative scenario there. However, leaving it does not mean categorically denying the possibility. History and biology have already demonstrated the error of assuming stasis in human affairs, while also teaching us the dangers of impatience and hubris. As of this writing, we still do not know whether such a suitable planet for what is called "terraforming" lies within our local stellar group, though that may be resolved in the next decade or soon thereafter. NASA currently has one mission, Kepler, dedicated to searching for exosolar planets that could become candidates for remote exploration in the distant future.

For those who wish a more detailed treatment of the many challenges facing our descendants for the next one thousand centuries, I recommend *Surviving 1,000 Centuries: Can We Do It?* by Roger-Maurice Bonnet and Lodewijk Woltjer.[5] This book treats a range of challenges people will face over this period, and offers the best summary of the asteroid-impact threat I have ever read. It makes a convincing argument why we should continue to pursue research leading to electricity generation by nuclear *fusion* power plants, which are *not* the same fuel-limited nuclear fission plants built today. It also gives strong arguments why the authors think it unlikely our descendants will leave earth during this period.

This is the only question where I take exception to the above authors' stated opinions. They make the case for the physical challenges very well. However, they seem to implicitly assume that the people living over the next thousand centuries will be very like us, will need to make the trip themselves, and may not have currently unavailable helpers that human ingenuity is starting to create, even today. In my opinion, this neglect of the future potential of human life and of some of our future creations is the reason we differ on this one speculative prediction. The future importance of further research on the still-biologically evolving human being and our own neural attributes not only has a direct impact on solving our most pressing current problems; it may also be the key to our descendants' distant future.

RETURNING TO THE PRESENT

However contemporary progressives and reactionaries may differ on almost everything else, there is one thing we all have in common. We all *live* in the present. So whether we look primarily forward or backward in our imaginations and our planning, we are all necessarily looking for solutions to contemporary problems. So were our progenitors, and so will be our descendants. In that sense, time is frozen for all of us. *For people to have a future, we must live in the present, and solve the problems of our present era.* This section, chapter, and book will end with an overview of what I think is most critical for our time.

If there is one thing I would improve rapidly and globally, it is education. I would give this the highest priority, even higher than working toward the solution of any of the other major problems discussed here. The reasons are many and simple but extremely compelling. First, ignorant people are easy to fool and easy to lead, and there are far too many individuals and institutions that would lead them into a harmful direction for personal or institutional gain.

The second reason follows largely from the first one. Without a larger fraction of the world's population—including the American population—

achieving a decent education, most of the great world problems discussed here cannot be solved in a timely manner. This is bound to lead to large-scale human suffering. History affords many examples of where it had to get worse before it started to get better. People do learn from suffering, but that is the hard way.

I distinguish as before between education and training. An educated person can think. Not all merely trained people can, or they have forgotten how, or they don't want to. Also, much of this education must be publically supported. Parochial education lends itself too easily to the biases of the supporting institution.

The education I describe here provides the knowledge that addresses and overcomes what I have called *ignorance of the first kind*. Provide that and some exercises in critical thinking—which can be satisfied by education in the nature of science—and the same great world problems that will otherwise long resist effective solutions could be solved more quickly. Human welfare will be enhanced. Human suffering will be reduced.

Research is the other activity I have supported through many chapters. This needs to be continued too, and the growing problem of better educating those who have become threatened by science when they do not like its conclusions must be addressed now. However, if an increase in budgets for both education and science were available, I would hold the research budget steady in real dollars and increase the one for quality public education. It does little good to produce a rapid increase in knowledge if most of the world's population has no idea what to do with the new knowledge or even opposes it. Research will continue, and some areas may need augmentation, while others may have to experience some lean years. I would favor more basic research on fundamental questions in the life sciences, including the neurosciences, but this is an issue for the political process to decide. Hopefully, those who place political leaders in their positions and the elected officials will make wise decisions.

The next major challenge I think should be met soon is addressing that stubborn structural problem of how society—which is rapidly becoming a global society—will manage mature multinational corporate capitalism. Most educated people have now recognized that this system, when unregulated, will run wildly out of control. Laissez-faire capitalists may deny this, but they are becoming increasing discredited by the actual performance of capitalism under pure free-market conditions, especially in its failure to provide socioeconomic justice.

I previously used the term *juggernaut technology* to describe this process, but upon reflection I now realize that the word *juggernaut* would probably have been better applied to this *under-regulated system of production*. It is not the technology itself but the way it is used that is the root problem. One of the most pernicious consequences of this system is its obsession with perpetual *growth*, to which the ruthlessly competitive, poorly regulated free-market process drives it. If this seems an extreme critique, it may seem less so in light of the ten major world problems identified by Jared Diamond in his previously mentioned book, *Collapse*.[6] "Growth" in most of the problem areas Diamond identifies will increasingly stress the human race. One of them, population increase, will drive most of the others. Perhaps it is necessary to remind people that not all forms of growth are necessarily benign.

Solutions to this problem have been offered, but none so far have worked well in practice. A mixed system of regulated capitalism and democratic socialism seems to have taken hold in the small, highly homogeneous Scandinavian countries, but time will tell if this continues to perform well. For reasons too numerous to review here, I am sure this would not work well in the United States today. Most scientists I know—including myself— would be comfortable working in a country under democratic socialism if it worked efficiently, but in most of the world today I am confident it would not. If some form of socialism is to be our economic future in a more advanced civilization, the human race needs to find a way to make a socialistic economy operate efficiently. That may be an even more difficult problem than learning how to tame capitalism without destroying its impressive capacity for innovation.

Before leaving a critique of inadequately regulated mature capitalism, I need to reemphasize how much the "captains of industry" are also trapped by this system. As long as they operate within the law, demonizing these people is a mistake. If they did not perform as they do, many of them would lose their jobs under this system as it is now. Nor should the workers be left entirely off the hook. Granting that they suffer much more from dips in a business cycle than the managers, the workers remain members of the society and must bear some responsibility for its performance. Anyone arguing that workers are helpless victims is arguing for a form of paternalism that no self-respecting worker should tolerate. In a free democratic society, workers can organize and they can vote. Hopefully, better education will help them understand what they are voting for.

Thus the problem of reining in mature corporate capitalism without destroying the innovative strengths of this system continues to be unsolved. Further education and research on this problem are needed. The solution is sure to come eventually. Meanwhile, we will move on, sometimes doing well, at other times collectively surviving and just getting by.

As for what is being badly neglected today, I would say ethics. Two examples will suffice to illustrate this. The first is the status of women in the world today. While women's rights vary greatly among nations and cultures, there is no major society today in which women receive full rights in practice, though in some they are accorded them in theory. Even in the United States, a leader in promoting women's rights, women are not accorded full rights in practice, as was demonstrated in chapter 6. In most of the Islamic countries, the situation is still appalling, even when allowance is made for historical and cultural idiosyncrasies. However, the Arab Spring suggests that this situation may be turning around, perhaps fairly quickly. Given that half of the brain power of the human race resides in women, there are compelling reasons not to waste it, but the real basis for equal rights for women is ethical.

The second example of overlooking ethics is the denial, obfuscation, exaggeration, and possibly outright lying to prevent political action to address human causation of climate change. Political discussions of this problem began over the science, moved on to well-funded efforts to confuse or distort the science, and have culminated in desperate attempts to discredit the scientists doing the research. Scientists and scholars are beginning to hit back, exposing not only the egregious arguments used to discredit the science, but also naming names and exposing the agendas of a small number of scientists engaged in the denial process, known collectively as "the climate contrarians."

This cultural counteroffensive has culminated in a growing recognition by the scholarly community that the root problem is an ethical one, the position taken in the political realm by Al Gore years ago. It is not ethical to misrepresent science, much less to lie about it, even if the person in question believes he is doing it to serve a higher purpose, such as preventing "creeping socialism."

In closing, we can note that the arts are sometimes more effective than the sciences in conveying even scientific and scholarly messages to large numbers of people. I attempted to do this once in the NASA of the early 1980s. The opportunity arose to produce a brochure describing a NASA mission for which I was then the project scientist. I wrote the text but wanted to include something that might provide that small shock to the general reader, which would hold his interest as he explored the technical passages. Mozart's opera *The Magic Flute* gave me what I needed.[7] By contemporary standards, some the social ideas of this opera are dated, at best. It contains recognizable sexist and racist passages. Nevertheless, it does honor the somewhat anachronistic, but then liberal, cult of the Freemasons; and the music was by Mozart. In the final scene, led by the leader of this cult, the chorus sings the praises of virtue and knowledge, and so on. I especially like the following words that I translated, loosely, from the German. Perhaps I found these words appropriate because my scientific specialty was solar physics.

> The Sun's golden rays pierce through the night,
> And scatter the powers of darkness to flight.

Light . . . that dissipates the darkness of ignorance and superstition, . . . that illuminates the world of nature and our own minds . . . that will eventually solve all of today's problems and guide our descendants to a brighter future.

> That was the Enlightenment. It still is, today.

NOTES AND BIBLIOGRAPHY

CHAPTER 1: THE HISTORICAL ENLIGHTENMENT

1. John Locke, *Concerning Civil Government, Second Essay*, in *Great Books of the Western World*, vol. 35, ed. Robert Maynard Hutchins (Chicago: Encyclopedia Britannica, 1952).

2. Voltaire, *Zadig, or Destiny*, in *Candide and Other Writings*, ed. and intro. by Haskell M. Block (New York: Modern Library, 1956). The translation in the reference is "He . . . knew as much of Metaphysics as has been known in any age, that is to say, very little."

3. John B. Bury, *The Idea of Progress* (New York: Dover, 1955). Both the author and historian Charles Beard in their two introductions emphasize that even the great thinkers of the classical world had no clear idea of progress, while the Christian theologians of the medieval West saw nothing but gloom and doom outside of salvation through the Church. Yet, as Bury points out, some of the first thinkers to see the new light were among the clergy, or at least were trained by them.

4. John Hale, *The Civilization of Europe in the Renaissance* (New York: Atheneum, 1994). See especially chap. 5, "Transformations," and the section titled "Unfrozen Voices," pp.189–215.

5. Isaiah Berlin, *The Proper Study of Mankind* (New York: Farrar, Straus and Giroux, 1998). See the section titled "The Originality of Machiavelli," pp. 269–326. Berlin suggests that Machiavelli gets a bad rap for his honest, if ruthless, advice in *The Prince*, which was written to save his native Italy from conquest by the large, better-organized powers to the north. Machiavelli lived when political heads were rolling everywhere and the brutal wars of the Reformation had not begun. In all other respects, he seems to have been an intelligent, sensitive man and a brilliant

diplomat. Berlin is not advising Machiavellian tactics today, or even condoning them then, but he exhibits a deep understanding of the man and his times. His writings exhibit the advice of many professional historians, to resist the temptation to judge the past by the standards of the present.

6. Edward Goodell, *The Noble Philosopher, Condorcet and the Enlightenment* (Amherst, NY: Prometheus Books, 1994). Condorcet may be an extreme example of a human being all of us would have admired, but his dreams for an ideal world were very premature for his own days and possibly for ours as well.

7. Francisco Goya, *Los Caprichos* (New York: Dover, 1969). These famous sketches excoriate human folly, especially those associated with superstition, including the famous "sleep of reason" mentioned in the main text that might have been written for the Enlightenment.

8. Juan José Junquera, *The Black Paintings of Goya* (London: Scala, 2003). Goya was Catholic, but he clearly did not favor religious fanaticism. His depiction of religious dementia in *The Saint Isidore Pilgrimage* is striking. Almost humorous is *The Sabbath*, in which a large goat performs a black mass. Different members of the audience are depicted as terrified, curious, or even indifferent. Goya had experienced an early enthusiasm for the Enlightenment, followed by the revolt against the French army in Spain, with numerous brutalities on both sides. The Black Paintings seem to have been his way of applying his motto "Ideal proportions be damned. I paint the truth." Goya had learned of ideal proportions in Italy, but he eventually went beyond them.

9. Bury, *Idea of Progress.*

10. Leslie A. White, *The Evolution of Culture: The Development of Civilization to the Fall of Rome* (Walnut Creek, CA: Left Coast Press, 2007). As one of the first anthropologists to adopt an evolutionary perspective in the field, White constantly makes the point that what really matters to all cultures is what works to promote two things: survival and reproduction. In this book, he attempts to roughly quantify how energy is put to work to promote these goals. Regardless of the value of this formulation, his approach is remarkably similar to the approach I have taken in this book. We regard the genetic nature of humankind (urging us to live and propagate) combined with reliable knowledge (science, because it works) as the key to future human survival and progress.

11. Lucretius, *On the Nature of Things*, in *Great Books of the Western World*, vol. 12 (see n. 1). As the standard treatment of Epicureanism today, this work is impressive for the remarkable speculations on the nature of the physical world. As an attempt at ethics, it is arguably a bit one-sided, making the case for letting the good times roll. For balance, the reader might consider Marcus Aurelius's *Meditations*, found in this

same volume and expressing a far more stoical outlook. Not surprisingly, Lucretius was apparently a wealthy Roman who could and did enjoy the good life. Marcus Aurelius was generally considered the greatest of the Roman emperors, whose life was unending responsibility that he seems to have discharged well.

12. Edward Gibbon, *The Decline and Fall of the Roman Empire,* in *Great Books of the Western World,* vol. 40 (see n. 1). Chapter 14 reviews the rise of Christianity in the empire and the conduct of the Roman government toward it, from Nero to Constantine. Chapter 28 discusses the collapse of paganism and the victory of the Christian religion in Rome, about which Gibbon seems to have had mixed feelings.

13. In 1957, Time Incorporated published *The World's Great Religions,* featuring contributions by many authors. The belief systems covered are Hinduism, Buddhism, Chinese philosophies, Islam, Judaism, and Christianity. The treatment is midlevel and colorful, and the similarities and differences are presented clearly. These accounts make clear that there are recognizable differences between the three East Asian and South Asian religious traditions, and between the three Western ones, if we assume that Islam, as an Abrahamic religion, is Western. Different attitudes of human beings toward nature are apparent. Perhaps the differences between "East and West" are diminishing today as the world slowly comes together. It is possible to imagine benefits to both.

14. Will Durant, *The Story of Philosophy* (New York: Simon and Schuster, 1953). Chapter 7 on Schopenhauer and, to a lesser degree, chapter 9 on Nietzsche review the work of two Western philosophers who were influenced by Eastern thought. Schopenhauer reflects the negative side of much Eastern thought with the view that the "blind will" of nature drives everything, but he notes the value of not-so-blind knowledge to manage it. Nietzsche's "eternal recurrence" is an interesting idea traceable to Hinduism: after a *very* long time the current state of affairs will repeat. Quantum physics has made that idea somewhat suspect. Both men were arguably misogynists. Neither wrote in the spirit of the Western Enlightenment, though Nietzsche did announce "the death of God," along with a prescient warning that some of the initial consequences might not be pleasant.

15. This is reference to Kipling's *Ballad of East and West,* where "never the twain shall meet." Though Kipling won a Nobel Prize in 1907, he is understandably out of fashion today, as an unabashed British Empire loyalist who spoke of "the white man's burden" and similar racist rant. He seemed to deny the possibility that East and West would ever come to a common understanding. There is now abundant evidence that Kipling, certainly no Enlightenment thinker, was wrong on that too . . . if we take the long view.

16. Bill Cooke, *A Wealth of Insights: Humanist Thought since the Enlightenment* (Amherst, NY: Prometheus Books, 2010). In contrast to the above three authors, Cooke *is* an Enlightenment enthusiast, and brings my chapter 1 up to date with a global perspective.

17. Francis Bacon, *Advancement of Learning, Novum Organum,* in *Great Books of the Western World,* vol. 30; Thomas Hobbes, *Leviathan,* in *Great Books of the Western World,* vol. 23 (see n. 1). Bacon was a lawyer who advocated experimental science without actually practicing it, but his work is often considered a major stimulant to the great British tradition of empiricism that so impressed Voltaire. Hobbes's argument for strong government begins with his assessment of human nature and culminates in his advocacy for the restraining influence of government and religion to manage a naturally somewhat barbaric humanity. His views are not entirely modern, but part of the dialogue Hobbes stimulated is still with us.

18. Locke, *Concerning Civil Government, Second Essay.*

19. David Hume, *An Inquiry concerning Human Understanding,* in *Great Books of the Western World,* vol. 35 (see n. 1). Hume was a gentle philosopher whose carefully reasoned skepticism may have demolished more sophisticated nonsense than that of any other thinker.

20. Adam Smith, *An Inquiry into the Nature and Causes of the Wealth of Nations,* in *Great Books of the Western World,* vol. 39 (see n. 1). Because Smith was a successful radical reformer of an overregulated economic system in his day (mercantilism), he continues to be the darling of the conservative unregulated free-market ideologues today.

21. Robert S. Alley, *James Madison on Religious Liberty* (Buffalo, NY: Prometheus Books, 1985). Thomas Jefferson regarded complete institutional separation of church and state as the single most important goal for the nascent United States, the one that would best guarantee the future success of democratic government. Alley reviews the role Jefferson's protégé James Madison played in facilitating this when both were alive. This study reviews one of Jefferson's last written requests to Madison: "Take care of me when I'm dead." To Madison, the meaning was clear. Some Americans today may not know this, and those few who openly prefer theocracy to democracy would certainly disagree.

22. Michel de Montaigne, *The Essays,* in *Great Books of the Western World,* vol. 25 (see n. 1). For moderation and good sense, these essays are hard to beat, especially considering that they were written when many Protestants and Catholics in Europe were killing one another with enthusiasm. Even on mortality, Montaigne's equanimity does not abandon him. He notes that dying requires no effort on our part. Nature takes care of it.

23. Jared Diamond, *Guns, Germs, and Steel—The Fates of Human Societies* (New York: W. W. Norton, 1997). Almost a modern version of an idea that Montesquieu (see note 24) advanced earlier, that geography plays a large role in societal destiny. The treatment is superb, and I have no quibble with the content, but I have a big one with the title, which should have been *Food, Germs, and Steel.* Anyone who has read this book will understand. Guns come last, *if* the more basic problems of survival have been solved. Was this a case of an editor choosing a "sexier" title?

24. Charles Montesquieu, *The Spirit of the Laws*, in *Great Books of the Western World*, vol. 38 (see n. 1). More on this author below, but his recognition of the importance of geography in co-determining a society's development is noteworthy.

25. Jean-Jacques Rousseau, *Émile, on Education* (New York: Basic Books, 1979).

26. Durant, *Story of Philosophy*, chap. 5, sect. 9. Voltaire's exact words, translated, when receiving Rousseau's *Discourse on the Origin of Inequality* were: "I have received sir, your book against the human race, and I thank you for it. . . . No one has ever been so witty as you are in trying to turn us into brutes; to read your book makes one long to go on all fours. As, however, it is now sixty years since I gave up the practice, I feel that it is unfortunately impossible for me to resume it." Language used as a rapier may have been risky then as now, but Voltaire was undoubtedly its master. Rousseau's essay largely condemns civilization and extols the virtuous "natural life," evocative of myths of an imagined earlier "golden age." Certain Europeans under Rousseau's influence speculated that the Indian inhabitants of North America might be such noble savages . . . until, like their European visitors, some proved less than noble.

27. Ibid., chap. 5, sect. 10. The procession described in the text was actually somewhat postponed, as Voltaire died in 1778 while the Old Regime in France was still in power. Voltaire was considered a terrible man, almost an atheist, by the conservatives, and his remains were first interred outside of Paris. In 1791, the National Assembly forced Louis XVI to return them to Paris, and that is when the procession took place. Even Durant, whose admiration for Voltaire knew few limits, admits he was a shameless self-promoter with an inordinate love of money. Yet arguably no single figure in history has done as much to limit the power of the Roman Catholic Church.

28. Philipp Blom, *A Wicked Company: The Forgotten Radicalism of the European Enlightenment* (New York: Basic Books, 2010).

29. Hugh Brogan, *Alexis de Tocqueville* (New Haven, CT, and London: Yale University Press, 2007). Tocqueville is probably better known in France for his study of *their* revolution, not the American democracy produced by ours. Among

his many insights was that revolutions occur not when people are hopeless but when enough of them see the possibility for improvement. A friend of reasonable liberals like Lafayette, Tocqueville achieved some prominence in French politics, but he did not live to see the Third Republic. His personal courage seems to have exceeded his constitutional strength.

30. Edmund Burke, *Reflections on the Revolution in France* (Harmondsworth, UK: Penguin Books, 1968). A sober assessment of when "revolution" attempts too much too soon and easily degenerates into tyranny. This remarkable conservative Whig supported the American Revolution in Parliament but fiercely opposed the more violent one in France for its attempt to remake humanity overnight and destroy all tradition. In the light of Burke's observations, it is hard not to reflect on the impact of academic and student radicals in America during the 1960s, and the subsequent middle-class reaction against it that began to flower under Reagan and culminated in the G. W. Bush administration, arguably one of the most reactionary in American history.

31. Henry Steele Commager, *The Empire of Reason: How Europe Imagined and America Realized the Enlightenment* (Garden City, NY: Anchor Press/Doubleday, 1977). This is excellent early American history, but final judgment might await future American generations deciding whether they want to honor the goals of the Enlightenment or try to rule the world in a spirit of American exceptionalism. Former Chinese premier Zhou Enlai is reputed to have assessed the impact of the American Revolution with "It's too early to tell." Having no regard for any Marxist dictatorship, I still find this observation compelling.

32. Bernard Bailyn, *The Ideological Origins of the American Revolution* (Cambridge, MA: Belknap Press of Harvard University Press, 1992). This treatment is slightly more nuanced than the reference in note 31. Bailyn gives due credit to the radical Enlightenment goals crafted largely in France but notes that the *political process* required to achieve these goals owes more to British political thought. This assessment is reflected in chapter 1 and also appears in the tradition of pragmatism in American philosophy, which may lack the elegance of the great British and Continental schools of empiricism and idealism but which places primary emphasis on "what works."

33. Henry Steele Commager and Richard B. Morris, eds., *The Spirit of 'Seventy-Six: The Story of the American Revolution as Told by Participants* (Indianapolis: Bobbs-Merrill, 1958). An outstanding collection of documents from both British and colonial American sources, revealing how controversial the American Revolution was in Great Britain itself, and what a near thing it was in the

future United States. In stages, the British realized that their policy toward the original colonies had been a mistake, and many Whig (liberal) politicians supported colonial independence in Parliament well before the British defeat at Yorktown. These same politicians became the leaders who devised the strategy that eventually defeated Napoleon. On the American side, Thomas Paine is among the authors featured in this collection. The first four sentences in Paine's *The American Crisis* helped inspire Washington's army during the harsh winter of 1776–1777, and Washington had the good sense to transmit these sentiments to the soldiers. They are worth repeating (see p. 505 in the reference for the direct quote):

> These are the times that try men's souls. The summer soldier and the sunshine patriot will, in this crisis, shrink from the service of their country; but he that stands it *now* deserves the love and thanks of man and woman. Tyranny, like hell, is not easily conquered; yet we have this consolation with us, that the harder the conflict, the more glorious the triumph. What we observe too cheap, we esteem too lightly . . . it would be strange indeed if so celestial an article as FREEDOM should not be highly rated.

(The italics and capitalization appear in the original.) The style is not modern, though the words seem to have made a difference when they were written. It would be a mistake to assume they were an encouragement to militarism, a confusion of means with ends.

34. Bailyn, *Ideological Origins.*

35. Thomas Paine, *Common Sense* (New York: Penguin American Library, 1982). Originally written in 1776, this tract went through fifty-six editions in that year alone. It was highly influential in generating support for the American Revolution in the year independence was declared.

36. Alexis de Tocqueville, *Democracy in America* (New York: Anchor Books/ Doubleday, 1969). Tocqueville's comments on the relationship he observed between church and state during his visit to America in 1831–1832 are expressed in book 1, chap. 9: "Most of English America was peopled by men who, having shaken off the pope's authority, acknowledged no other religious supremacy; they therefore brought to the New World a Christianity which I can only describe as democratic and republican; this fact singularly favored the establishment of a temporal republic and democracy." Regarding the Catholic clergy at that time, Tocqueville makes the following statement in the same chapter: "American priests proclaim themselves in general terms in favor of civil liberties. . . . *They are at pains to keep out of affairs and not mix in the combination of parties*" (the italics are mine).

37. Brogan, *Alexis de Tocqueville*. When some gifted future author writes the history of our imperial phase, it is comforting to think that *Democracy in America* may find its way into some future set of classics.

38. Hale, *Civilization of Europe*.

39. Paul Kurtz, *In Defense of Secular Humanism* (Buffalo, NY: Prometheus Books, 1983). Some contemporary atheists have decided that secular humanism is passé. This book, written by a distinguished professor of philosophy and the founder of a major international secular humanist organization, reviews the intense efforts of political and religious reactionaries in the twentieth century to discredit and marginalize secular humanism. It is not difficult to predict that when angry atheism has run its course (again), it will be the continuation of a sober-minded, scientifically based secular humanism that will continue the progress toward a more progressive world based on science and humanist ethics . . . in effect, on the broad goals of the Enlightenment.

40. Alan Bullock, *The Humanist Tradition in the West* (New York: W. W. Norton, 1985). The broad-mindedness of this history is reflected in the following statement that appears on p. 160. "While accepting that the humanist tradition inherits from the eighteenth and nineteenth centuries a current of anti-Christian feeling as one of its historical characteristics, the claim sometimes made by both secularists and fundamentalists that secularism *represents* humanism is a travesty—as much a travesty as to take fundamentalism to represent religion." Sobering is the final chapter, chap. 5, titled "Has Humanism a Future?" Bullock mentions four contemporary concerns that he feels threaten humanism, based on the years 1933–1953. They are, as he numbers them: (1) The growth of population and the altered scale of history. (2) Technology and the altered pace of history. (3) Collectivism and the totalitarian state. (4) Wars and the spread of violence. Have things changed that much since 1953?

CHAPTER 2: STATUS OF ENLIGHTENMENT GOALS TODAY

1. Naomi Klein, *The Shock Doctrine: The Rise of Disaster Capitalism* (New York: Picador, 2007). This reference gives an up-to-date critique of inadequately regulated mature capitalism in the contemporary global free market. The argument is convincing that while the wealthy prosper, the economic impact on most

others has been negative. The critique of the Milton Friedman school of free-market modeling is particularly good. The mathematically sophisticated models produced by that school are based on the classical assumption that today's consumers are rational buyers, a highly questionable assumption in the present day of psychologically astute marketing and political sound bites.

2. Four excellent recent books review the tactics of those who oppose government regulation of greenhouse gases to minimize climate change. These are: (1) David Michaels, *Doubt Is Their Product: How Industry's Assault on Science Threatens Your Health* (Oxford: Oxford University Press, 2008). The tactics that worked for the American tobacco industry have been applied more recently to block action that could reduce human-made climate change. Because science does not produce absolute certainty, doubt is deliberately created in the minds of a gullible public to prevent regulation. (2) Gwynne Dyer, *Climate Wars: The Fight for Survival as the World Overheats* (Oxford, UK: Oneworld, 2010). This is a geopolitical analysis of the many global disasters, including military ones, likely to follow a failure to rein in current energy-company efforts to prevent action to reduce global warming. (3) Naomi Oreskes and Erik M. Conway, *Merchants of Doubt: How a Handful of Scientists Obscured the Truth on Issues from Tobacco Smoke to Global Warming* (New York: Bloomsbury, 2010). This work brings David Michaels's book (above) up to date on the impact of deliberate deception on the climate-change issue. (4) Stephen M. Gardiner, *A Perfect Moral Storm: The Ethical Tragedy of Climate Change* (Oxford: Oxford University Press, 2011). The latest in this series makes a strong ethics-based case against deliberate misrepresentation of the science to prevent action on industrial activity causing climate change. Gardiner takes up the ethical theme that was emphasized earlier by Al Gore and that appears in Gore's book, *Our Choice, A Plan to Solve the Climate Crisis* (Emmaus, PA: Rondale, 2009).

3. *Bulletin of the Atomic Scientists* (published by Sage, http://www.journals@ sagepub.com) has consistently emphasized the danger of actually using nuclear weapons, the immense destructive capacity of which may now far exceed the public imagination, due to familiarization. When any American administration considers a possible first strike using such a weapon, the public should be concerned. This publication can be counted on to raise the alarm. Biosecurity, climate change, and nuclear-power generation are other subjects reviewed in this journal.

4. Steven Pinker, *The Better Angels of Our Nature: Why Violence Has Declined* (New York: Penguin, 2011). This book provides an optimistic picture of how violence is declining and offers abundant statistical proof that the mortality rate due to violence has noticeably declined over the past millennium in much of the world.

However, the case becomes weaker when the increase in the global population is factored in and absolute numbers are used. Most disturbing to Americans should be fig. 3-10, which shows that homicide rates in the United States at the end of the twentieth century are as high as they were at the beginning. Nevertheless, Pinker's arguments suggest that more people may be waking up to the harmful consequences of authoritarianism, war, ignorance, poverty, mendacity, and greed, and may be willing to do something about it. This is a hopeful book motivated by a decent outlook.

5. *Climate Change 2007, The Physical Science Basis* (New York: Cambridge University Press, 2007). This is the Science Working Group's full report resulting from the Fourth Assessment of the Intergovernmental Panel on Climate Change (IPCC), the recognized gold standard on climate science through 2005. (The fifth report is not yet available.) Climate change contrarians often divert attention to a short summary for policymakers, suggesting that the IPCC effort is political, and conveniently ignoring that this much more extensive volume is the summary of the *science* done by approximately two thousand researchers with expertise in climate science. The report summarizes the work of climate scientists who publish most of the papers that appear in the refereed literature on that subject. Many of the contrarians duck that challenge, preferring to issue politically oriented, unrefereed publications instead.

6. See note 2 for the titles in question. While these studies deal specifically with the attempts to discredit legitimate science in support of economic and/or political agendas, they also emphasize how destructive pseudoscientific arguments are in confusing a vulnerable public on the nature of science itself. Since scientific knowledge is our most reliable source of information on everything natural, discrediting science undermines the knowledge upon which solutions to an increasing number of major world problems must be based.

7. Kendrick Frazier, ed., *Science under Siege* (Amherst, NY: Prometheus Books, 2009). A collection of articles reviewing how those threatened by science have been mounting increasingly well-funded campaigns to either deny legitimate science, as in the case of so-called creation science, or to substitute a bogus so-called sound science to discredit climate science. Authors who contributed to this book include Ken Frazier, Chris Mooney, Carl Sagan, Mario Bunge, Joe Nickell, Barbara Scott, Martin Gardner, Steven Pinker, David Morrison, and Stuart Jordan.

8. Carl Coon, *One Planet, One People* (Amherst, NY: Prometheus Books, 2004). A former US ambassador with experience in the Middle East and South Asia argues why humanity has become one already, and why this requires a political

evolution to future global institutions that have the authority and the resources to deal with major world problems that lie beyond the ability of any one nation to solve. This may sound—and may be—unrealistic today, but the case for the need is made clearly.

CHAPTER 3: IGNORANCE, SUPERSTITION, AND JUGGERNAUT TECHNOLOGY

1. Ronald A. Lindsay, *Future Bioethics, Overcoming Taboos, Myths, and Dogmas* (Amherst, NY: Prometheus Books, 2008). In chapter 7, Lindsay reviews the illogical, unscientific, even absurd arguments offered by religious reactionaries to prevent the use of embryonic stem cells for medical research.

2. Eugenie C. Scott and Glenn Branch, eds., *Not in Our Classrooms: Why Intelligent Design Is Wrong for Our Schools* (Boston: Beacon Press Books, 2006). As an example of why so much American public science education is failing, Scott, Branch, and other authors discuss the unscientific, ideologically based reasons behind the attempt to teach "creationism" and describe how to combat this pernicious tendency. Barbara Forrest and Paul R. Gross, *Creationism's Trojan Horse: The Wedge of Intelligent Design* (Oxford: Oxford University Press, 2007). This edition includes a chapter on the Dover, Pennsylvania, trial and provides an interesting example of how the "creation science" community keeps changing its strategy to undercut legitimate biological science in the service of its religious ideology. The latest attempt that postdates this book seems to be an argument to "teach the controversy." There is hardly any scientific controversy, only a major political one.

3. The definition of *superstition* given in this chapter is a direct quote from *The American Collegiate Dictionary* (New York: Random House, 1969).

4. Thomas Hobbes, *Leviathan*, in *Great Books of the Western World*, vol. 23, ed. Robert Maynard Hutchins (Chicago: Encyclopedia Britannica, 1952). Hobbes's conclusion that people need a powerful authoritarian ruler to protect all from the ravages of all sounds pessimistic and undemocratic to us today, and it is. However, Hobbes's argument is well made, and if "the people" are unable to support a more liberal society, history suggests that the "Leviathan" is exactly what they will get. In the context of this chapter, we may reasonably conclude that under earlier and more difficult conditions of life, patriarchal authoritarian rule may have been almost inevitable for large communities under duress. Enlightenment thought

proposes that through science and reason we can and should get beyond that. At the moment, it would seem we are still trying.

5. The description of the Juggernaut in the text is taken from *The Encyclopedia Britannica*, vol. 13 (New York, London, and Toronto, 1959), p. 175.

6. Marshall McLuhan is famous for the catchy phrase "The medium is the message." McLuhan might be called the first, most prescient thinker to discern the triumph of technology over rational discourse. If you perform a Google® search for this phrase, don't be surprised to see "The medium is the massage." It's the same message.

7. Jacques Ellul, *Propaganda: The Formation of Men's Attitudes* (New York: Vintage Books, 1973). The first edition was published in 1965. This is the finest description of modern propaganda I have ever read. It goes beyond Joseph Goebbels's formulations created to inculcate ideas into people's heads without their conscious knowledge to recognizing that propaganda and the modern technological society are inextricably bound together, and the one cannot thrive without the other. (Not surprisingly, the author also wrote a book titled *The Technological Society*.) Ellul argues that propaganda is a deadly enemy of democracy, because the entire purpose is subliminal conversion to the advertiser's or the politician's purpose. Modern marketing that puts the lie to the flattering (and easily sold) idea of the rational consumer is only one example. Certain forms of political messaging in the United States and elsewhere today may be even more pernicious, but those who craft the messages are often the same people well versed in exploiting human psychological vulnerabilities.

8. To learn how costs for "doing business in space" rose during the early stages of the NASA Space Shuttle Program, go to the available archives of studies funded by the now defunct (unfunded) US Congress Office of Technology Assessment (OTA). The OTA's site lists the sequence at http://www.princeton.edu/~ota/disk3/1982/8226/822610.PDF (appendix D) and http://www.princeton.edu/~ota/disk3/1982/8226/822611.PDF (appendix E). Note how, according to appendix D, it seemed that the shuttle was going to prove less expensive than a Delta Rocket launch (but with growing expense estimates with time), and the following, alarming, and ultimately prescient, staggeringly high estimates for shuttle deployment made by James Van Allen, discoverer of the magnetospheric radiation belt that bears his name today. I was a NASA scientist at that time, and I knew no working NASA engineer who thought the shuttle would be cost-effective for unmanned satellite launches. The reason was transparent. To safely "man rate" a shuttle flight in orbit is very expensive. Just making sure there are no sharp edges that might rip a secure suit during an EVA is not "cheap." And there is more.

CHAPTER 4: THE CRITICAL ROLE OF SCIENCE

1. Two excellent books on the life of Albert Einstein illustrate the role of imagination in the life of the most famous physicist of the twentieth century. Jürgen Neffe, *Einstein: A Biography* (New York: Farrar, Straus and Giroux, 2005) and Walter Isaacson, *Einstein: His Life and Universe* (New York: Simon and Schuster, 2007) give remarkably similar accounts of the life and work of this remarkable scientist; both works are based on the recent availability to scholars of the "Cal Tech papers" on Einstein's life. For excellent descriptions of how Einstein developed the physical insights that led to relativity and other still-remarkable achievements in 1905, see chap. 7 in the Neffe reference. For the development of special relativity in particular, see chap. 6 in Isaacson.

2. Andrew D. White, *A History of the Warfare of Science with Theology in Christendom* (New York: Dover, 1955). The Dover edition is a reprint of a book originally published in 1895 that remains a classic to this day. White is less hostile toward religion than the title suggests. He does not write in the spirit of "the new atheism" that tends to attack all religion, tolerant or otherwise. However, he brilliantly excoriates all attempts of religion to pontificate on the workings of nature, providing literally hundreds of examples of how those who get their science from ancient religious texts today are made to look pathetically foolish and ignorant. Fortunately, most of the *tolerant* religious would find little in White to offend them today.

3. Lauri Lebo, *The Devil in Dover: An Insider's Story of Dogma v. Darwin in Small-Town America* (New York: New Press, 2008). The book was selected as the first recipient of the Studs and Ida Terkel Author Fund to support New Press authors, and it won high praise from Pulitzer Prize–winning author Edward Humes. Lebo's account of the Dover trial should alert the major media that there remain principled, courageous people in the hinterland who will not submit to the current journalistic fashion to give equal weight to both sides of an alleged controversy, when one side presents carefully reasoned arguments based on evidence and the other is left with rationalizations based on claims of ancient texts and possible lies in court. For an account of the trial coauthored by the current head of the National Center for Science Education, see chap. 10 in Barbara Forrest and Paul R. Gross, *Creationism's Trojan Horse: The Wedge of Intelligent Design* (New York: Oxford University Press, 2007). Be sure to get the 2007 edition that includes chap. 10.

4. Fyodor Dostoyevsky, *The Brothers Karamazov* (New York: Modern Library, 1950). The scholarly Ivan Karamazov is troubled by thinking morality may fail without a God and ultimately suffers a nervous breakdown. The deformed brother, Smerdyakov, takes the idea seriously and murders their tyrannical father. Ivan also writes a disturbing essay that deconstructs the Christian religion by postulating a secret meeting between a captive Jesus and "the Grand Inquisitor," a Torquemada who explains how the Church has learned to manage people through "miracle, mystery, and authority." The brilliant Czarist-era novel raises several questions that still trouble many sincere religious people. Secular humanists have resolved these questions for themselves in the Enlightenment tradition.

5. Sigmund Freud, *Civilization and Its Discontents*, in *Great Books of the Western World*, vol. 54, ed. Robert Maynard Hutchins (Chicago: Encyclopedia Britannica, 1952). Often regarded as the finest essay ever written by Freud, the work argues that human assertiveness, human aggression—often the product of frustrated assertiveness—and human sexuality do not fit easily into the requirements of modern civilization. This view fits with evolutionary biology, which demonstrates that we evolved biologically under harsher conditions. Thus, along with cultural co-determinants of our behavior, these earlier-evolved genetic co-determinants continue to influence our actions. Freud notes that the management of natural impulses is extremely challenging. From this, we might conclude that achieving "the perfectibility of man" is a far more difficult undertaking than some of the optimistic Enlightenment thinkers and certain scholars have since hoped it would be, something modern history seems to have confirmed. Contemporary neuroscience is beginning to suggest that the problem could be more acute for the male of our species, who may be more inclined *by nature* to be violently aggressive and sexually promiscuous, and also less empathetic than the female. While still controversial and sometimes categorically denied, this has profound implications for our descendants' future if it is true.

6. That Freud's above thesis (though not necessarily many of his psychoanalytic speculations) needs to taken seriously is reinforced by many observations in *The Lessons of History*, written by Will and Ariel Durant (New York: Simon and Schuster, 1968). The Durants produced the last five volumes of their well-known eleven-volume *Story of Civilization* together and summarized their main conclusions in the book referenced here.

7. Immanuel Kant, *The Critique of Pure Reason* and *The Critique of Practical Reason*, in *Great Books of the Western World*, vol. 42 (see n. 5). I would not presume to grasp the subtle points of this distinguished philosopher, and the references given here are mainly to point to the sequence in which they appeared, which

was in the years 1781 and 1788, respectively. That sequence is significant. Kant wished to establish, among other things, an indestructible basis for the "metaphysics of morals," a term he used frequently. This brilliant logician may not have been entirely satisfied with the first work, as it affected a practical approach to morality. The second work is sometimes regarded as his attempt to do that. Today, there seems to be a general consensus that, in spite of Kant's superb and extremely complex arguments, in which he advanced our thinking on morality significantly, he nevertheless failed to achieve the higher goal he initially had in mind. Thus ethics and moral philosophy continue to be fruitful fields for further study. Several major modern questions suggest they are fields badly in need of examination by a continuously developing field of ethics, in which philosophers work with the sciences where relevant scientific knowledge is available, which—it is important to note—is often still not the case today.

8. Edward O. Wilson, *Consilience: The Unity of Knowledge* (New York: Alfred A. Knopf, 1998). The somewhat-unfamiliar title is based on an idea developed in the nineteenth century. *Consilience* is defined in the second paragraph of chap. 2 as "the linking of facts and fact-based theory across disciplines to create a common groundwork of explanation." Thus, taken to the limit where all knowledge of the natural world has become available and there is no supernatural order that influences the natural world, everything in our natural world is not only explainable, but the explanations are all connected in a self-consistent way. All features of human behavior and culture become understandable on fundamental grounds that ultimately depend on the illusive "theory of everything" that theoretical physics and cosmology seek but have not yet discovered.

While no thinking person today would claim that humanity is even close to approaching this consilience of all knowledge, the concept has great motivating power among scientists, many of whom are atheists. Yet for those among the religious who look with disdain on this idea, they would do well to read the brief, five-page chap. 1 of this book. They would be hard-pressed to find a minister of any gospel more forthright, more modest, or more facile in expression than this author. Wilson is a world authority in contemporary biology who has won two Pulitzer Prizes for describing his work to a larger public. Yet, having once been a Christian before studying Darwin, he grasps the yearning to understand the world shared by all people, including the religious. Wilson has reached out to Evangelicals in America to join the scientific community in the struggle to preserve the global ecosystem against the devastation caused by unregulated exploitation. Some in that community are listening.

CHAPTER 5: SCIENCE FOR LIFE

1. Steven Pinker, *The Blank Slate: The Modern Denial of Human Nature* (New York: Viking Press, 2002). Pinker provides a needed antidote to the extreme views of those who deny human nature in the interest of postmodernist or deconstructionist ideologies. These were popular until recently in many liberal arts departments in the West. Perhaps this was a reaction to the horrors of Nazi racism based on terrible biology. Some of these intellectuals were Jewish, the people who arguably suffered the most. That might explain a lot of anti-biological thinking, which fortunately seems to be dying out today.

2. Daniel C. Dennett, *Darwin's Dangerous Idea: Evolution and the Meanings of Life* (New York: Simon & Schuster, 1995). Of the so-called four horsemen of atheism (Richard Dawkins, Dennett, Sam Harris, and Christopher Hitchens), this secular author may exhibit in some of his other work the most subtle and nuanced understanding of why so many intelligent people continue to support religion today (see note 12). However, Dennett arguably makes a more powerful argument against "supernaturalism" than any of the other authors. There is a reason why he calls the Darwinian paradigm "royal acid": its power to dissolve false concepts. No wonder some creationists attack the Darwinian paradigm. They grasp the implications that Dennett discusses well in this book.

3. Ernst Mayr, *The Growth of Biological Thought, Diversity, Evolution, and Inheritance* (Cambridge, MA: Belknap, 1982). This is a superb treatment of the "New Synthesis" in evolutionary biology, for a reader who wants to understand how it was developed. The book was written by one of the main developers.

4. Barbara Forrest and Paul Gross, *Creationism's Trojan Horse: The Wedge of Intelligent Design* (New York: Oxford University Press, 2007). Chapters 8 and 9 are especially good for revealing how the clearly religious motives for sneaking pseudoscientific "intelligent design" into public schools have been implemented.

5. Daniel C. Dennett, *Consciousness Explained* (New York: Little, Brown, 1991). A fine overview of what is grasped about the phenomenon—some might say the epiphenomenon—of our awareness, and so forth, in general terms. However, any reader expecting a claim that consciousness is understood in detail by the exacting standards of neuroscience will be disappointed, as I'm guessing Dennett would grant. Thus, my only quibble with this book is its title. For an easier treatment that shows how subtle Dennett is, try his *Elbow Room: The Varieties of Free Will Worth Wanting* (Cambridge, MA: MIT Press, 1993). Does Dennett really mean free will as defined by philosophers or else commonly understood? The reader must decide.

6. Jesse M. Bering, "The Cognitive Psychology of Belief in the Supernatural," *American Scientist* 94, no. 2 (2006).

7. Shaun Nichols, "Experimental Philosophy and the Problem of Free Will," *Science* 331 (2011).

8. Tom Wolfe and Michael Gazzaniga, "Free Will," in *Science Is Culture: Conversations at the New Intersection of Science + Society*, ed. Adam Bly (New York: Harper Perennial, 2010). This conversation reveals differences between a well-known author's perception of free will and that of a scientist who has made major contributions to cognitive neuroscience. The more cautious (conservative?) Gazzaniga is uncomfortable with attributing the "essence" of a person to brain states alone. Wolfe is more accepting of that. An interesting dialogue.

9. Jaak Panksepp, "Empathy and the Laws of Affect," *Science* 334 (2011). A succinct review of contemporary research on sources of empathy in the brain, including work on mirror neurons, which permit us to grasp certain mental states of ourselves in others. The article is written by an expert in the field who has also explored sources of rage behavior in the brain, work that influenced Steven Pinker's book *The Better Angels of Our Nature*, which is discussed in chap. 4. The research on mirror neurons may offer hope of finding ways to enhance empathy among humans, short of performing deliberate changes in the human genome for which we are clearly not ready today.

10. Ray Kurzweil, *The Singularity Is Near: When Humans Transcend Biology* (London: Penguin, 2005). Kurzweil is a transhumanist who seems to look forward to transcending the limitations of organic life. I must grant his ability for great achievement in the world of fast computers, but I wonder if he adequately appreciates the achievements and future potential of organic life.

11. John Hatcher, *The Black Death: A Personal History* (Philadelphia: Da Capo, 2008). By "personal," the author is referring to the impact of the bubonic plague on the inhabitants of one village in East Anglia that suffered the passage of the deadly illness in the fourteenth century. The town lost over one-third of its people in a few spring months, then recovered surprisingly quickly. Survivors did the only thing possible. They resumed their lives as they had been before, though with different people who replaced those they had recently lost. The author is a world authority on the most lethal event in European history, and he bases his account as much as possible on actual records (usually kept by surviving clergy). While his account probably cannot penetrate the psychological suffering of the survivors, it is a remarkable tribute to human toughness in a crisis. Perhaps the Ingmar Bergman film *The Seventh Seal* offers a hint why. The stoic knight recently

returned from the Crusades does not survive his chess match with "Death" (a monk). A silly young couple almost oblivious to the horror around them does. It would seem that this simple will to live and propagate is central to human nature.

12. Daniel C. Dennett, *Breaking the Spell* (New York: Penguin, 2007). This is Dennett's contribution to the four books by the "four horsemen" of the alleged New Atheism, which created some stir when they first appeared. Dennett demonstrates his usual subtlety by exhibiting a secular humanist's understanding of some of the reasons why many people, including a number of well-educated people, continue to support religion. Here, he advances the idea of "believing in belief" and recognizes that religion continues to provide part of the societal "glue" that holds society together. This prompts two comments. First, believing in belief is not necessarily the same thing as belief itself. Second, this view is similar to attitudes shared by many philosophically inclined thinkers of the Enlightenment, including some atheists, reflecting how slowly things often change on a large societal scale.

13. Thomas Mann, *The Magic Mountain* (as *Der Zauberberg*; Berlin: Alfred A. Knopf, 1927). In Mann's novel, the brilliant but dogmatic Catholic priest Napta shoots himself in rage after his Enlightenment opponent, Settembrini, fires his weapon into the air during a duel that resolves their competition for the mind of a young tubercular engineer. Literary critics may see other things in this, but to me it is an imaginative demonstration of what can happen to a brilliant but dogmatic mind when its neural house of cards crafted from impeccable logic collapses. The reference to "not with a whimper but a bang" reverses the concluding line of T. S. Eliot's poem *The Hollow Men*, which ends "this is the way the world ends," followed by the final phrase "not with a bang but a whimper." Presumably the poet, who was deeply religious, considered many people to be superficial. We can grant that Napta was not superficial, but we may still prefer Settembrini.

14. John Rawls, *A Theory of Justice: Revised Edition* (Cambridge, MA: Belknap Press of Harvard University Press, 2003). While the complete work is a monumental tome, chap. 1, "Justice as Fairness," covers only the first 47 of 517 pages and summarizes the process for which Rawls is famous.

15. Jacques Ellul, *Propaganda* (New York: Vintage Books, 1973). The author learned much about propaganda during World War II from the Nazi occupation of his country, but his insights into how advertisers use propaganda for marketing in a technological society reveal many similarities of technique.

16. Kurzweil, *Singularity Is Near*. The remarks in the text are an attempt to be fair to the views of a brilliant computer scientist who may not understand the biological world as well as the inorganic one. He might say that the same laws of

physics apply in both cases. True. But do they apply in the same way? Are adaptive processes in biological organisms that are associated with their neural functions completely compatible with inorganic machines, however "smart"?

17. Peter Snow, "Woe, Superman?" *Oxford Today* 22, no. 1 (2009).

18. Erich Fromm, *The Anatomy of Human Destructiveness* (New York: Holt Paperback, 1992). This book on aggression by a well-known mid-twentieth-century social psychologist was once popular. However, comparing its very general categorizations of aggressive types with some of the modern work on this subject based on neuroscience shows how far we have come in half a century in advancing toward a more scientific approach to this critical subject. Fromm's treatment is very broad and general, and somewhat speculative.

CHAPTER 6: UNIVERSAL HUMAN RIGHTS

1. John Rawls, *A Theory of Justice: Revised Edition* (Cambridge, MA: Belknap Press of Harvard University Press, 2003).

2. Leslie A. White, *The Evolution of Culture: The Development of Civilization to the Fall of Rome* (New York: McGraw-Hill, 1959).

3. Bertold Spuler, *History of the Mongols* (New York: Dorset, 1988).

4. Steven Pinker, *The Better Angels of Our Nature: Why Violence Has Declined* (New York: Viking, 2011).

5. Lynn Hunt, *Inventing Human Rights* (New York: Norton, 2007).

6. P. Gassier and J. Wilson, *The Life and Complete Works of Francisco Goya* (New York: Harrison House, 1981). This collection offers many examples of three characteristics of Goya's work, especially his later work: He is sometimes described as the first great Western artist to exhibit sympathy for ordinary working people and the poor, the source of his popularity with many Marxist writers. He excoriates the mendacity and dullness of many of the Catholic clergy and much of the Spanish aristocracy during his lifetime, though he remained a nominal Catholic and did some fine works of art for that church, while exhibiting in his satirical sketches a profound contempt for superstition in general. Goya finally rejected the conventional classical styles he had studied in Italy and created an original one based on, in his own words, "Velázquez, Rembrandt, and Nature." French cultural critic André Malraux noted, "With Goya, modern art begins."

7. Nassim Nicholas Taleb, *The Black Swan: The Impact of the Highly Improbable*

(New York: Random House, 2010). The title says it all. The "black swan" is a stand-in for any event that is very rare and almost impossible to predict yet that has major consequences, typically harmful. An example is the recent near collapse of the American economy, for which there were (as always) a few who predicted it, either from deep insights or from lucky guesses. It amused some of us around Y2K to read about stockbrokers who predicted endless economic prosperity, while others said the 1990s must eventually end. Likewise we noted how many Republicans argued that the huge surplus the government had amassed should lead to slashing taxes radically, while equally optimistic Democrats were busy crafting new government programs to spend all that extra money. I found Taleb's argument about "black swan" events to be very convincing. (The author was a successful stockbroker for years and seems to have left that profession in time to avoid the recent "readjustment.") I found only one idea Taleb expresses and defends that I find wrong. The subsection titled "History Does Not Crawl, It Jumps" states this idea near the beginning of the book and develops it in many succeeding chapters. History does jump (right). But it also crawls (a serious error to deny it). Without periods of near stasis to consolidate things, we would probably not be here at all, that is, we would have no history at all. Once again the Darwinian paradigm provides a useful model to understand this. Biological life-forms—which is what we are—are unlikely to survive if they experience nothing but an endless succession of "punctuated equilibriums" (biological black swans) that are usually harmful to most extant species. The times between these dramatic events are what we need to catch our collective breath until the next black-swan event hits. Nevertheless, I found this to be a fascinating book that presents an important idea very well.

8. Steven Bach, *Leni: The Life and Work of Leni Riefenstahl* (New York: Vintage Books, 2008). The relevance of this biography to the Enlightenment is how it describes the antithesis of the Enlightenment, namely passionate patriarchal political romanticism. Leni Riefenstahl is an outstanding example of a highly talented woman of humble beginnings who clearly succumbed to an extreme form of romanticism in both her personal life and in the political life of her country. Given certain features of the German culture in which she grew up, this is understandable, but the political consequences of her propaganda films are disturbing, for they helped Hitler and Goebbels present a heroic image of Nazism to their people that made them even more formidable in World War II. It is somewhat reassuring that *this* particular propaganda may not have worked in the America of my youth, thanks to the ending of *Triumph of the Will*, in which German soldiers goose step through Nuremberg, and stick-figure Heinrich Himmler "Heil Hitler's" in a way

that once provoked raucous laughter here. (Himmler may have looked like a fool, but the German Army was no joke, possibly thanks in part to Riefenstahl's very effective propaganda.) An attempt was made to make an American movie on Leni's life, but this never came to fruition, which is arguably unfortunate. Leni declared that Jodie Foster, who was to play the leading role, was "not beautiful enough." I doubt that. Apparently the two met, and Leni may have realized that Jodie Foster probably understood "Leni" all too well. This is an excellent biography to read, if you want an insight into romanticism running amok in politics, and the propaganda this can produce.

9. Nell I. Painter, *The History of White People* (New York: W. W. Norton, 2010).

10. William W. Freehling, *The Road to Disunion: Secessionists at Bay, 1776–1854* (New York: Oxford University Press, 1990).

11. Ida B. Wells-Barnett, *On Lynchings* (Amherst, NY: Humanity Books, 2002).

12. Thomas Dixon Jr., *The Clansman* (New York: Grosset & Dunlap, 1905). Subtitled "An Historical Romance of the Ku Klux Klan." No. I'm not joking. Hardly a classic, it nonetheless provided D. W. Griffith with the plot for one of the most famous silent movies ever made, *Birth of a Nation*. When the Ku Klux Klan rides to save the honor of a presumably threatened white woman to the background music of Wagner's *Ride of the Valkyries*, it is hard to keep a straight face. Klansmen were not ghosts; it's history that haunts us.

13. The account of the Nat Turner rebellion is based on the "Nat Turner" entry in *Encyclopedia Britannica*, vol. 22 (Chicago: Encyclopedia Britannica, 1959), p. 628.

14. Norm Allen, ed., *African American Humanism: An Anthology* (Amherst, NY: Prometheus Books, 1991).

15. Norm Allen, ed., *The Black Humanist Experience: An Alternative to Religion* (Amherst, NY: Prometheus Books, 2003).

16. Robert W. Fogel and Stanley L. Engerman, *Time on the Cross: The Economics of American Negro Slavery* (New York: W. W. Norton, 1989). The surprise in this book is the conclusion about the decent diet that was available to many of the slaves. Otherwise, the institution was generally as bad as most of us now realize, which is also noted in this study.

17. R. E. Nisbett and D. Cohen, *Culture of Honor: The Psychology of Violence in the (American) South* (New York: Harper & Collins, 1996). The experimenters placed an associate in a narrow hallway, with instructions to bump against numbers of exclusively Northern young men and other Northern young men who had spent substantial time in the South, after which the testosterone levels in the

two groups were taken and averaged. The men who had not lived in the South exhibited normal testosterone levels. Those who had absorbed some Southern culture had elevated testosterone levels, exhibited higher levels of a known stress hormone, and scored lower on a standardized test to measure self-esteem. You do not have to be a sociologist to know that American Southern culture is a more traditional "honor culture" than is the northern part of the country. And the South is arguably proud of that.

18. Richard M. Restak, *The Brain: The Last Frontier* (New York: Warner Books, 1979). While somewhat dated, this book (and its author) were extremely popular when first published, for it introduced many of us who are not neuroscientists to this now rapidly growing field. Chapter 10 in particular discusses some of the early research that established definite neuroanatomical (structural) differences between the (statistically averaged) brains of men and women, especially the larger corpus callosum in women's brains that seems to facilitate more cross talk between the two hemispheres. More recent research has extended to an attempt to determine what statistically significant neurophysiological (functional) differences there might be as well. This has produced a cottage industry of speculation about why men and women and societies are the way they are, and why society should either change, stay the same, or regress to an earlier stage of dominant patriarchy. The subject could hardly be more controversial, except among educated seculars, who, with rare exceptions, now strongly support equal rights for women. Unfortunately, not everyone is onboard yet.

19. At the height of the movement to improve women's rights in the United States during the 1970s, several serious studies of the status of women were written by women and were widely read. Betty Friedan's *The Feminine Mystique* (New York: Dell, 1963) may have been the most analytical, but many other authors made the same point. It was argued that the popular image of women was largely a product of men's ideas of the kind of woman they wanted, and most women had been working hard to conform to it. On the European side, and also widely read here, was Simone de Beauvoir's *The Second Sex* (New York: Bantam Books, 1952). More personal than Friedan's book but also analytical, Beauvoir's work examines some common myths about women held by five well-known authors, among whom the best known to Americans are probably D. H. Lawrence and Stendhal. Beauvoir understood how important Enlightenment ideals are to the ongoing movement to liberate women economically and to involve them politically. More current and indicative of both progressive tendencies and current restraints in American politics is a refreshing account of *Sex, Science, and Stem Cells: Inside the Right Wing Assault on Reason*, by Colorado congresswoman Diana DeGette with Daniel

Paisner (Guilford, CT: Lyons Press, 2008). I cannot refrain from quoting one of her choice descriptions of conservative men in the House of Representatives, and I hope the quote, if read, will help, not harm, her political career.

> Over time, I realized that the politicization of science by the Republicans and the religious right was at its most insidious over any issue relating to human reproduction. This brought me to the inevitable conclusion that too many of our elected officials are simply incapable of thinking rationally about sex. . . . The disconnect was so transparent that some of our older male politicians couldn't even talk about any aspect of human sexuality without biting their lips to avoid snickering like school boys. There was a general and widespread discomfort in talking about sex, except in the form of bawdy jokes.

These remarks appear in chap. 1, titled "Where I Stand." I think we need more women in Congress, and, just as the marines might add, more *good* men. But then no one can entirely escape the culture of his youth.

20. Richard Dawkins, *The Selfish Gene (New Edition)* (New York: Oxford University Press, 1989). See especially chap. 11, "Memes: The New Replicators," in which Dawkins gives a succinct description of the dynamic relationship between gene and meme and how together they co-determine who and what we are. Like many initial skeptics, I was at first unimpressed with the meme concept, finding it to be merely a new term for already-established powerful cultural concepts such as "God"—in effect, pouring old wine into a new bottle. When it finally sunk in how parallel the meme is to the gene as the cultural twin to the genetic co-determinant, I changed my mind. In particular, both function powerfully in silence, including the meme when it operates on the subliminal level, which it often does. This idea can be applied to how religion is often taught by clergy. Religious teachers who have been long saturated in a dogmatic religious culture could easily be infected with the God "virus of the mind" without realizing it. They could then unwittingly transmit it to those not yet exposed. This raises the question of how well these people fully understand what they are doing, and why. Of course, this can be said about the rest of us engaged in other endeavors too. To live in a culture and be influenced by it automatically removes the possibility of complete objectivity, and most of us prefer that to social isolation that can easily induce insanity. I now find the meme concept a useful aid to thought. Dawkins's book is an excellent presentation of both "selfish gene and subtle meme."

21. Alexander Werth, *Russia at War* (New York: Carroll & Graf, 1964). If this seems an odd reference for a chapter on human rights, think what *personal* impact it must have had on people living in even a victorious country to have lost almost one-sixth of its entire population over four years, independent of the brutal political system that directed its efforts. Werth's account lists "only" twenty million deaths. That was the official account given by the communist regime when Werth's book was written. When the USSR collapsed, the real official records were released. That number was twenty-eight million. If men in authoritarian societies are the prime movers and actors in large-scale warfare, we need to recognize it and learn how to prevent it. It is hard to think that engaging more smart women on this problem would not help us find increasingly humane solutions.

CHAPTER 7: RELIGION EVOLVING

1. John Hatcher, *The Black Death* (Philadelphia: Da Capo, 2008). One of the interesting conclusions Hatcher reaches after studying the records from a village in East Anglia that may have lost 40 percent of its people in a few months concerns the impact of the plague on the "ordinary" people as opposed to the clergy. Each group seems to have responded according to what one might have expected. The clergy seemed even more shocked and troubled by this once-mysterious and fatal menace but continued to show compassion to the people. The surviving people simply carried on as before and seemed to exhibit more down-to-earth common sense. The gods do what they do, but meanwhile people have to eat.

2. Alfred F. Loisy, *The Gospel and the Church* (Buffalo, NY: Prometheus Books, 1988). A classic of biblical criticism with an introduction by biblical scholar R. Joseph Hoffmann, this book gives the views of a Catholic priest who advocated modern tolerant views on religion that the upper levels of the Church hierarchy found deeply threatening to their more traditional interpretation of Catholicism. Loisy was among "the Modernists" who opposed this traditional doctrine. That stimulated a strong reaction from the higher ecclesiastical hierarchy, leading to the promulgation of a papal encyclical issued in 1907 by Pope Pius X. This study contains the complete text of that encyclical, which is titled "On the Doctrine of the Modernists." It is nothing less than a medieval-minded frontal assault on the Enlightenment and almost everything that movement stands for. If the contemporary Catholic Church values democracy, among other things, reading this encyc-

lical today should embarrass them. It reeks of authoritarianism and may explain one reason why author John Cromwell felt compelled to write *Hitler's Pope* about the later Pius XII (New York: Viking, 1999). Pius XII seemed sympathetic to the Axis dictatorships, though there were *some* mitigating circumstances. Communism was no friend of Catholicism, and the Germans were combating the USSR during the war. There had also been a moderate encyclical condemning anti-Semitism issued by Pope Pius XI in 1937. Unfortunately, his successor, Pius XII, was more accommodating to the Axis. The moderate encyclical seems to have had no effect. Loisy was excommunicated for his troubles.

3. Karen Armstrong, *A History of God* (New York: Ballantine Books, 1993). This is where Armstrong describes the "Axial Age," in which the major world religions, especially those in the current Western and South Asian world assumed many of the characteristics they still retain today.

4. Leslie A. White, *The Evolution of Culture: The Development of Civilization to the Fall of Rome* (New York: McGraw-Hill, 1959). See especially chap. 13, "The State Church: Its Forms and Functions," in which this anthropologist notes that "we should expect to find that the functions of the church, and the role that it plays in the social organism, are fundamentally like those of the state. This, as a matter of fact, is what we do find." These words should be no surprise to any thinking person actively engaged in society today. In spite of this, many highly intelligent atheists find it hard to grasp how difficult it is going to be to disentangle government from religion, when religion remains central to how many people still think. Not surprisingly, nontheists tend to divide into two groups: those who continue to reach out to the tolerant religious and others who find them oppressive. What was true of church-state relations during earlier stages of civilization clearly remains true today, certainly among Americans.

5. Jean-Marie Chauvet, Éliette Brunel Deschamps, and Christian Hillaire, *Dawn of Art: The Chauvet Cave (The Oldest Known Paintings in the World)* (London: Thames & Hudson, 1996). Both the account of the discovery of this cave and the magnificent thirty-thousand-year-old paintings of the animals (many large and dangerous) are impressive. Interpretations of why they were painted with such skill vary, but it is hard not to think some ancient preceptor would not have pointed to these paintings during a "lecture" to children and said, "If you see something like this out there, get back into the cave. Fast!"

6. Terence McKenna, *Food of the Gods: Search for the Original Tree of Knowledge* (New York: Bantam Books, 1993). The author provides arguments speculating that early shamans might have been the first humans convinced of a "spirit world"

while under the influence of hallucinogenic substances obtained from plants. This sounds familiar to anyone who observed the popularity of drugs during the 1960s, and it probably resonates with those who use them today. The approach is not so much that of the scholar as of one genuinely interested in the powerful effect drugs can have on the imagination. The author seems to regret the loss of the magical world of the mushroom, and he does treat some possible advantages when the need arises for inducing composure. The author's case that hallucinogenic drugs may have contributed to early beliefs in a "spirit world" is credible.

7. Among the many recent books written to defend atheism and, in some cases, to attack all gods if not all religions, are *The God Delusion* by biologist Richard Dawkins (Boston: Houghton Mifflin, 2008) and *Breaking the Spell* by philosopher of science Daniel C. Dennett (London: Penguin Books, 2006). What is noteworthy about these two books is the complete agreement between the two distinguished authors on the atheist epistemology, combined with how they differ in their subtle interpretations of the societal value or lack of value of tolerant religion. Dawkins is more hard-core and finds little to recommend religion, basing this on the usual (valid) arguments that there is no shred of evidence for the truth of any dogmatic religious claim, and noting documented harms committed in religion's name. You have to admire his forthrightness, but is this the right approach for society today? Atheist Dennett hints that it might not be, noting that while many outwardly religious people share at least the extreme skepticism—or the atheism—of the nontheist, they think "believing in belief" may be a good thing. This is based partly on the thought that human civilization, overall, may not be ready for atheism and that believing in belief remains part of the "glue" that still holds society together. That is not an original idea with Dennett; it was popular among deists and probably some atheist Enlightenment figures during the eighteenth century. However, it is interesting to see a modern atheist express this idea. Could this reflect a difference between two similar but subtly different intellectual traditions—one British and the other a very multicultural American one?

8. Dennett, *Breaking the Spell*.

9. Dawkins, *God Delusion*.

10. Samuel Eliot Morison, *The Oxford History of the American People* (New York: Oxford University Press, 1965). See especially chap. 10, sect. 4, "The Great Awakening in Religion." This was probably the most radical of America's three eruptions of religious enthusiasm, of which the third seems to still be with us. Morison is a distinguished historian with a decidedly conservative outlook. He

is clearly sympathetic to religion in its societal role, as other chapters in this book demonstrate.

11. Jeffrey Toobin, *The Nine: Inside the Secret World of the Supreme Court* (New York: Doubleday, 2007). The book makes a strong case that the Court is less inclined to legislate than is commonly believed and instead reflects the spirit of the times, whether liberal or conservative. That is somewhat reassuring.

CHAPTER 8: THE PRIMACY OF POLITICS TODAY

1. John B. Bury, *The Idea of Progress* (New York: Dover, 1955).

2. Clark Wissler, *Indians of the United States* (Garden City, NY: Doubleday Anchor Books, 1940). Chap. 11, "The Iroquois Family," which discusses the Cherokee nation, is especially good. The Iroquois influenced early New England in many ways, not only in their democratic tendencies. Of course, things were not always peaceful on either side. The Cherokee enjoy the distinction of having produced their own written language. Largely uprooted by President Andrew Jackson and moved to "Indian Territory" (present-day Oklahoma), some of them subsequently fought for the Confederacy in the American Civil War. Their history is told well in John Ehle's *Trail of Tears: The Rise and Fall of the Cherokee Nation* (Garden City, NY: Doubleday Anchor Books, 1988). The quaint view many contemporary Euro Americans still hold of the Native Americans is shown to be generally well intended, but patronizing.

3. Jefferson Davis, *The Rise and Fall of the Confederate Government*, 2 vols. (New York: Da Capo, 1990). The original was published in 1881. There is an excellent introduction by James McPherson, whose Pulitzer Prize–winning account of the Civil War, *Battle Cry of Freedom* (New York: Oxford University Press, 1988), exhibits little sympathy for the Southern culture of that era. Nevertheless, McPherson grants that Davis's arguments for secession, reviewed in vol. 1, were rational . . . and had also been advanced by certain Northern states before 1860 in the service of different causes. The usual way of solving such otherwise unsolvable problems was then invoked, and both authors describe the military action from their different perspectives. A frequent view of professional American historians is that the Civil War, more than the earlier Revolution, should be viewed as the defining event in American history, by beginning the expansion of human rights the Revolution promised but could not deliver. The relevance to the Enlightenment

should be transparent. The relevance to the present effort of certain political and religious extremists to subvert this expansion and drag the society backward to a more patrician era should be equally obvious.

4. Paul Kengor, *The Crusader: Ronald Reagan and the Fall of Communism* (New York: Harper Collins, 2006). Those progressives who refuse to examine evidence that politicians they opposed may have achieved a major success will not like this book. Granting that it is a shameless hagiography of Ronald Reagan, the case the author makes that Reagan was single-mindedly dedicated to bringing down the former USSR—and succeeded in doing so before bureaucratic strangulation did so naturally—is very strong. Perhaps the best argument left for those categorically opposed to Reagan is that he was playing a very dangerous game, based on the assumption that the Soviet leaders of the 1980s, for all their bluster, would not behave like the proverbial cornered rat and destroy civilization, with considerable help from the United States. That is a progressive position worth sober reflection. One thing is obvious. Due in part to this major success, Reaganism has become a cult of the political right in America today, in economics as well as foreign affairs. One of the clear consequences is the extreme libertarian mantra that since the Soviets were crumbling from too much government regulation, we should go to the other extreme and eliminate government regulation almost entirely. After all, President Reagan did say "The problem is the government." But would the USSR have collapsed in 1990 in any event? This book suggests that the correct answer is probably not.

5. David Niose, "No Agenda? A Humanistic View of Justice Scalia," *Humanist* 70, no. 2 (2010). The author is an attorney, and served as president of the AHA when the article appeared. Niose's brief study of Justice Scalia makes a strong case that this distinguished jurist has exhibited a proclivity for slanting decisions in the direction of traditional Catholic teachings.

6. John Rawls, *A Theory of Justice: Revised Edition* (Cambridge, MA: Belknap Press of Harvard University Press, 1991).

7. The publication *Defense Monitor* offers a good example of criticism of what *many high-level professional military officers* have considered waste and excess in certain defense appropriations. A good example appears in vol. 39, no. 2 (2010). The full program to build the new F-35 Strike Fighter has now been costed out to approximately one trillion dollars to completion. Some features of the planned aircraft are arguably either so advanced there is no foreseeable enemy that requires these capabilities, or, even more disturbing, they may be difficult to achieve anyway. I lack the competence to judge the many claims, but I will say that (1) I know the

defense prime contractor to be highly competent (i.e., that's not the issue), and (2) you can bet that you will see many full-page ads in American newspapers, and the like, explaining why the United States "must have" this capability as soon as possible. *Defense Monitor* is published by the World Security Institute's Center for Defense Information in Washington, DC.

8. Barbara W. Tuchman, *The Proud Tower: A Portrait of the World before the War: 1890–1914* (New York: Macmillan, 1966). This is a superb account of the arrogance of *every* major Western nation before the first "great war" of the early twentieth century.

9. Isaiah Berlin, *The Proper Study of Mankind* (New York: Farrar, Straus and Giroux, 1998). See the selection "Herder and the Enlightenment" in the part titled "History of Ideas."

10. Steven Pinker, *The Better Angels of Our Nature: Why Violence Has Declined* (New York: Viking, 2011). See chap. 9 in the section titled "Empathy."

11. David Bergamini, *Japan's Imperial Conspiracy* (New York: Pocket Books, 1972). Bergamini's work has been criticized extensively, yet the once-popular idea that Emperor Hirohito stoutly opposed what Americans call "the Pacific War" that Bergamini first exposed as convenient political fiction—for America as well as for Japan—is now widely accepted. (Did people foolishly expect that a Japanese emperor would oppose his own people in war?) More relevant to this chapter is Bergamini's discussion of a strong Japanese democratic movement in the 1920s. This failed in part due to economic hardship during the global depression and also from "government by assassination," arranged primarily by army militarists.

12. John W. Dower, *Embracing Defeat: Japan in the Wake of World War II* (New York: W. W. Norton, 1999). Winner of both a Pulitzer Prize and a National Book Award, this is an excellent study of the depressing conditions in a nation that rose rapidly in the modern world, crashed wretchedly, and then stumbled erratically for a few years to find a better way than rampant militarism. The first years after World War II were turbulent and depressing, and the recovery to a prosperous, more democratic society after that was relatively rapid. The book ends as that recovery begins.

13. Samuel P. Huntington, *The Clash of Civilizations and the Remaking of World Order* (New York: Simon & Schuster, 1996); also Joseph S. Nye Jr., *The Paradox of American Power* (New York: Oxford University Press, 2002). It is tempting to call these two foreign-policy experts "the dueling Harvard policy wonks," though the former is now deceased and the latter formally retired (but still active). Both distinguished scholars, Huntington was the more conservative on this issue. Nye's

views on "soft power," elaborated in many books, certainly conform more to the Enlightenment outlook, but Huntington makes a strong case to be prepared for the worst, with allies. Both are worth reading, perhaps together.

14. Stuart Jordan, "How We Can Win the Global Culture War," *Free Inquiry* 25, no. 2 (2006). This article resulted from a conversation I had with the founder of the Center for Inquiry, Paul Kurtz, who requested it. This book resulted from that article, which was a brief discussion of some ideas that appear here. I would have preferred a somewhat less dramatic title.

15. Alexis de Tocqueville, *Democracy in America* (New York: Harper and Row, 1966).

16. Robert Boston, *Why the Religious Right Is Wrong about Separation of Church and State* (Amherst, NY: Prometheus Books, 1993). The author gives a historical account that reveals how poorly church-state separation has fared in America, from Protestant domination of public education in the mid-nineteenth century (chap. 4) to the partially successful effort to introduce "parochiaid" recently (chap. 6). See also Edd Doerr and Albert J. Menendez, *Church Schools & Public Money: The Politics of Parochiaid* (Amherst, NY: Prometheus Books, 1991). These two authors show that in all states but one where voters were asked if they approved tax support for religious schools, the majority rejected the measure and Catholic voters aligned with Protestants in rejecting it. Not surprisingly, Protestant fundamentalists and the higher Catholic clergy have been the most active advocates of extracting public monies to support parochial schools.

17. Carl Coon, *Culture Wars and the Global Village: A Diplomat's Perspective* (Amherst, NY: Prometheus Books, 2000) and Carl Coon, *One Planet, One People: Beyond "Us vs. Them"* (Amherst, NY: Prometheus Books, 2004).

18. Jared Diamond, *Collapse: How Societies Choose to Fail or Succeed* (New York: Penguin Books, 2005).

CHAPTER 9: PLANETARY HUMANISM

1. Austin Dacey, *The Secular Conscience: Why Belief Belongs in Public Life* (Amherst, NY: Prometheus Books, 2008). Chap. 3 provides a particularly good discussion of Spinoza's role in the developments of ethics. Dacey notes there that "from a purely pragmatic standpoint, secularists should avail themselves of the arguments that reach the religious where they live—at the level of conscience." I

am not convinced that applies to the intolerant religious, but it certainly applies to the tolerant among them.

2. John Stuart Mill, *On Liberty*, in *Great Books of the Western World*, vol. 43, ed. Robert Maynard Hutchins (Chicago: Encyclopedia Britannica, 1952). Mill offers the classical argument for placing essentially no restraints on free speech or writing. Some contemporary interpretations are controversial, such as the view that corporations should have the same "free speech" rights as individuals in America today. Mill lived in an earlier economic era and while sometimes called a libertarian and a utilitarian, he might be expected to question some of those positions today. This is one of the best-argued essays against suppression of free expression ever written.

3. John Rawls, *Theory of Justice: Revised Edition* (Cambridge, MA: Belknap Press of Harvard University, 1999).

4. Carl Coon, *One Planet, One People: Beyond "Us vs. Them"* (Amherst, NY: Prometheus Books, 2004).

5. Carl Coon, *Culture Wars and the Global Village* (Amherst, NY: Prometheus Books, 2000).

6. Ronald A. Lindsay, *Future Bioethics: Overcoming Taboos, Myths, and Dogmas* (Amherst, NY: Prometheus Books, 2008). The arguments are all presented logically, starting with the established facts in each case, and reaching easily understood, sensible conclusions. Readers might find chap. 3, "Legalizing Physician Assistance in Dying for the Terminally Ill," particularly interesting. Even those categorically opposed to such measures might be impressed with the well-argued case that concerns for the often-claimed "slippery slope" are greatly exaggerated, based on evidence from where the practice is already legal.

7. Paul Kurtz, *Forbidden Fruit: The Ethics of Secular Humanism* (Amherst, NY: Prometheus Books, 2008). This is a reprint of a popular common-sense guide to secular ethics, with a new prologue to this 2008 edition. Kurtz emphasizes the common moral decencies that are universal among all surviving major cultures, though modes of application and punishments for infringements can and do differ. This book has been used in discussions of secular ethics by many local secular humanist communities, including one with which I am associated.

8. Stuart Jordan, "The Ethics of Secular Humanism," in *Developing Sanity in Human Affairs*, ed. Susan Kodish and Robert Holston (Westport, CT: Greenwood). Based on leading a year-long discussion of the kind mentioned in note 7, above, I was asked to review our conclusions at Hofstra University during a conference organized by professionals in general semantics. Our series began with Kurtz's ideas and then developed ten societal guidelines that, applied together, would implement

the secular ethics described in *Forbidden Fruit*. The critical feature of the resulting synthesis was the need for all ten guidelines to be applied together. For example, the guideline of liberty was defined simply as the freedom to do as one pleased without interfering with an equal freedom of anyone else. Any rational person can see this would be pragmatically inadequate without certain other guidelines, in particular, justice. No subset of guidelines was considered complete without compromising the proper application of at least one of the others. The approach of seeking a logical interconnection among the ten guidelines to provide a complete set that might provide the basis for an ethical society was the main feature of the talk/paper. Though the approach was deliberately very "grassroots" and simple, it was well received in a rather academic setting. The merit of this approach, inspired by Kurtz's work, is that it is easily grasped by any intelligent person, though it can be based on more academic ethical philosophy.

9. George Antonius, *The Arab Awakening: The Story of the Arab National Movement* (Safety Harbor, FL: Simon, 1939). The Wahabi movement is discussed in detail along with the general state of the Arab world on the eve of World War II. It may not have been the recent Arab Spring, but Antonius shows that things were stirring in the Arab world well before the twentieth century. The reason we call it Saudi Arabia under Wahabi influence today, and not Feisali Arabia, is interesting. (The Wahabi fellow was a better general in fighting the Turks.) Also insightful is Antonius's view of the Balfour Declaration, which may have been issued primarily to further British wartime and imperial policy during and after World War I more than to favor European Jews. The British and the French after World War I seem to have made a mess of the Middle East, and it's hard to escape the impression that, since then, we Americans have walked right into it. That must probably be left to the experts, but which ones? A somewhat different and more recent view of the Arab world was written by a distinguished Princeton historian who for a while advised the G. W. Bush administration. In 2002, Bernard Lewis published *What Went Wrong? Western Impact and Middle East Response* (Oxford: Oxford University Press). Lewis's simple but convincing thesis is that since the Mongols destroyed old Babylon, the more conservative clerics have gradually taken over, ending the great days of Arab culture that slipped further and further behind a reenergized West, until some Muslims finally woke up and realized how relatively backward they had become. Lewis uses a very Western term to describe what he thinks is needed, a *Reformation*. That may be true, but I would have preferred a more neutral term for Muslim ears. *Reformation* is no doubt a less annoying term in this context than *Crusade*, but it still sounds like a projection of European thinking. However, Lewis's

arguments for reform are very convincing. Perhaps the Arab Spring will bring that reform. I like Lewis's implicit—but perhaps not explicit—idea that ultimately it is the people of the Arab world and no one else who should decide their future. Isn't that real democracy? Attacks *from* that world are another matter.

CHAPTER 10: OVERCOMING CRIPPLING IGNORANCE

1. See chap. 2, note 2 for the five relevant references.

2. Yudhijit Bhattacharjee, "Eugenie Scott Toils in Defense of Evolution," *Science* 324, no. 5932 (June, 5, 2009): 1250–51. As a longtime director of the National Center for Science Education when this interview took place, Scott describes succinctly how the forces of the anti-Darwin movement in America are themselves "evolving." Their strategies change like the colors of a chameleon once their latest bit of nonsense has been exposed. We wait with baited breath for what will replace "teach the (scientifically nonexistent) controversy" once that is exposed. Unfortunately, something probably will.

3. Paul R. Gross and Norman Levitt, *Higher Superstition* (Baltimore: Johns Hopkins University Press, 1997). A marvelous debunking of the idea popular in some deconstructionist circles that science is just as arbitrary as any other human attempt to understand the world. The authors effectively deconstruct the work of those deconstructionists, thus "elucidating sources of bias in their text."

4. Susan Jacoby, *The Age of American Unreason* (New York: Pantheon Books, 2008). Chap. 10, titled "Junk Thought," is particularly good. Not surprisingly, what certain lobbyists call "sound science" often proves to be junk thought. Lobbyists with an agenda that differs from science learn to play the word game well.

5. Jared Diamond, *Collapse: How Societies Choose to Fail or Succeed* (New York: Penguin Books, 2005). The message that a people must adapt to changing geographical conditions or else fail is becoming increasingly relevant as climate change accelerates. The inability of a northern European people to learn from the Inuit in Greenland is particularly instructive. The Norwegians are famous for their exploits at the poles of the earth, but their colony on Greenland eventually perished.

6. Naomi Klein, *The Shock Doctrine: The Rise of Disaster Capitalism* (New York: Picador, 2007). When the Occupy Wall Street movement came to New York, this author was there with them.

7. John Stuart Mill, *On Liberty*, in *Great Books of the Western World*, vol. 43, ed. Robert Maynard Hutchins (Chicago: Encyclopedia Britannica, 1952). It is hard to think that Mill would not have been pleased to see free speech in action during the Occupy Wall Street activities. How many Americans know that Mill was not an academic but a high-ranking official of the British East India Company, with acute financial judgment?

8. Bhattacharjee, "Eugenie Scott Toils." The popularity of creationism may be the best evidence for how poor general science education is in America. In spite of impressive achievements by many American scientists, standardized tests suggest that the country ranks near the bottom among developed countries in educating a larger population in the sciences. See note 2.

9. Stuart Jordan, "Defending Climate Science Today," *Humanist* 70, no. 4 (2010). The article emphasizes how, as soon as one proposed natural mechanism for global warming has been shown not to work, the climate contrarian community quickly proposes another to keep climate scientists busy. *No mechanism studied to date, other than anthropogenic greenhouse gases, has achieved scientific credibility.*

10. *Climate Change 2007: The Physical Science Basis*, Fourth Assessment Report of the Intergovernmental Panel on Climate Change (New York: Cambridge University Press, 2007). Called the IPCC-2007 Science Report, it has been carefully scrutinized, as it should have been, and an error was discovered in a reported estimate of how quickly the Himalayan glaciers that provide water for much of northern India might melt away. That became a favorite critique of climate-change contrarians, who ignored an otherwise-enormous body of evidence that supported human causation of contemporary climate change and withstood scrutiny. Recent research has even increased the IPCC-2007 estimate of sea-level rise in the current century by a factor of three, should new measurements of Greenland icecap melting persist. See also http://www.tandfonline.com/doi/abs/10.1080/037 36245.2009.9725337.

11. Judit M. Pap and Peter Fox, eds., *Solar Variability and Its Effects on Climate—Geophysical Monograph 141* (Washington, DC: American Geophysical Union, 2004). Readers of this book are encouraged to see if the very people—solar physicists—who would have received more grants by showing solar causation of recent global warming did in fact present such a case. Not even significant correlations between phases in the solar cycle and global terrestrial temperatures could be found, even allowing for phases shifts. Yet, one still hears the contrarian claim "The Sun has produced the recent global warming."

12. Jordan, "Defending Climate Science Today."

CHAPTER 11: REASSESSING THE ENLIGHTENMENT TODAY

1. This comment was delivered by the anthropologist Lionel Tiger during an unpublished talk given at the New York Academy of Sciences in 1999, sponsored by the Council of Secular Humanism. The obvious connection between the beliefs of a simple herding and agricultural people and major myths that retain living force today provoked a serious discussion, perhaps the highlight of this rather scholarly gathering.

2. Naomi Klein, *The Shock Doctrine: The Rise of Disaster Capitalism* (New York: Picador, 2007).

3. Daniel Kahneman, *Thinking, Fast and Slow* (New York: Farrar, Straus and Giroux, 2011).

4. Nassim Nicholas Taleb, *The Black Swan: The Impact of the Highly Improbable* (New York: Random House, 2010).

5. David Warsh, *Knowledge and the Wealth of Nations: A Story of Economic Discovery* (New York: W. W. Norton, 2006).

6. Klein, *Shock Doctrine*.

7. Paul Kengor, *The Crusader* (New York: HarperCollins, 2006).

8. Chris Hedges, *American Fascists: The Christian Right and the War on America* (New York: Free Press, 2006). The best part of this book may be the short introductory essay by Umberto Eco titled "Eternal Fascism: Fourteen Ways of Looking at a Black Shirt." This title alone suggests Eco's familiarity with his subject. The fourteen characteristics of fascism described by Eco are a cult of tradition; a rejection of modernism; regarding distinctions to be a sign of modernism; action for action's sake; opposing diversity of political views; general individual and social frustration; chauvinistic nationalism; popular humiliation by the wealth and power of others; belief that life is lived for struggle; contempt for the weak; aspirations to heroism; a preference for mastery and power in sexual matters; selective populism (among the followers of a "Common Will"); and deliberately speaking with George Orwell's simplistic Newspeak to restrict critical thinking. By noting similarities with *some* of these defining characteristics, Hedges provides examples to argue that the American government in power when his book was written may have been moving in a reactionary fascist direction.

9. Isaiah Berlin, *The Proper Study of Mankind* (New York: Farrar, Straus and Giroux, 1998). Berlin argues against historical inevitability in a section by that

title. That would seem to be against the general idea presented here that scientific knowledge combined with enduring qualities of human nature make it very likely our more distant descendants will be around and will live in a more progressive civilization than we have today. If Berlin's arguments for skepticism on this are applied on a time scale of a few centuries, I would embrace his position on *major* societal progress for reasons given throughout this book. It is impossible to predict that humanity will avoid further large-scale disasters, given what we know today. However, it seems likely that the stubborn will to live is likely to override collectively suicidal behavior and that these disasters too, if they occur, will be overcome. That leaves our more remote descendants in a much better position to achieve major societal progress if scientific knowledge continues to increase, especially the arguably most important kind, which is more reliable knowledge of ourselves. I find the question of time scale critical in assessing the matter of whether or not our descendants can get beyond "muddling through" to a demonstrably higher stage of civilization. On a relatively short time scale, I would say no, simply because we are still individually and collectively too ignorant. That would agree with Berlin, who may have been reacting to the disasters of the two world wars of the early twentieth century. However, looking ahead, say, a thousand years or more, the odds look increasingly better. The thesis of this book is, with no apology, a *mild form* of scientism. It is not a claim that adequate, reliable scientific knowledge is available now, or soon will be, to realize the kind of society some of the more optimistic of the Enlightenment thinkers thought would emerge relatively quickly. Condorcet did not grasp the staggering complexity and difficulty of the problem. Isaiah Berlin did.

10. The project in St. Louis that was demolished only a few decades after completion was Pruitt-Igoe. It quickly degenerated into a crime-infested slum.

11. Philipp Blom, *A Wicked Company: The Forgotten Radicalism of the European Enlightenment* (New York: Basic Books, 2010).

12. Jared Diamond, *Collapse: How Societies Choose to Fail or Succeed* (New York: Penguin Books, 2005).

13. Voltaire, *Candide,* in *Candide and Other Writings,* Modern Library ed. (New York: Random House, 1956).

CHAPTER 12: TOWARD A BETTER FUTURE

1. Edward O. Wilson, *Consilience: The Unity of Knowledge* (New York: Alfred A. Knopf, 1998).

2. Sigmund Freud, *Civilization and Its Discontents,* in *Great Books of the Western World,* vol. 54, ed. Robert Maynard Hutchins (Chicago: Encyclopedia Britannica, 1952).

3. Roger-Maurice Bonnet, and Lodewijk Woltjer, *Surviving 1,000 Centuries: Can We Do It?* (Berlin: Springer Praxis, 2008).

4. Jean-Marie Chauvet, Éliette Brunel Deschamps, and Christian Hillaire, *Dawn of Art: The Chauvet Cave (The Oldest Known Paintings in the World)* (London: Thames & Hudson, 1996).

5. Bonnet and Woltjer, *Surviving 1,000 Centuries.*

6. Jared Diamond, *Collapse: How Societies Choose to Fail or Succeed* (New York: Penguin Books, 2005).

7. Wolfgang A. Mozart, composer, and Emanuel Schikaneder, librettist, *The Magic Flute* (in German, *Die Zauberflöte*). Premiered in 1791.

ACKNOWLEDGMENTS

Few books are the ideas of one person, and this one certainly fits that pattern. There are many persons I should thank in the course of writing the book, and it is inevitable that I will overlook someone who should be mentioned and regret it later. To minimize that possibility, I will proceed chronologically. The first person to suggest that I might undertake this project was the one to whom the book is dedicated. It is with sorrow I must say that Paul Kurtz saw neither the text nor the dedication that I wrote two years ago when the writing commenced. I had hoped it would offer him my appreciation for his achievements for the American—and the international—secular humanist movement, but Paul died suddenly on October 20 of this year. At a dinner held in Washington, DC, during December 2005, Paul read a short article I had written, asked that I submit it to the *Free Inquiry* magazine, and later suggested that it might be expanded into a book. I was too busy to consider this at the time, but in the summer of 2008 finally decided that, yes, I would like to do that.

Four people then had to suffer through reading a very quick summary of many of the ideas that appear here. They are, in alphabetical order, former ambassador Carl Coon, anthropologist and university administrator Dr. Nancie Gonzalez, intrepid microbiologist from the upper peninsula of Michigan Sandra Hubscher, and English teacher and bibliophile Dr. Ruth Mitchell. Their feedback was extremely valuable, both encouraging and critical. I was especially amused, and appreciative, of one critique. "You have some valuable ideas here, but until this is organized it reads like a brain dump!" Since the subject of the book attempts to tie together disparate threads, all in support of a central thesis about the Enlightenment, the reader must decide if I have succeed in the organizational task. For those familiar with the term, this is an attempt to achieve "the view from nowhere," which is actually the view from everywhere that is relevant to the central issue.

The next step involved preparing a proposal for a book, which was handled by Steven L. Mitchell, editor in chief of Prometheus Books. Steven issued clear instructions for preparation and scheduling from the start, and in the final preparation of the book I am grateful for the professionalism of Catherine Roberts-Abel, Julia DeGraf, and Jade Zora Ballard. All of these people and other members of the staff at Prometheus Books permitted me as author to concentrate on the writing while they worried about production matters.

Beyond these people are many others whose conversations over the years have contributed greatly to the content. Colleagues in science at the Goddard Space Flight Center are so numerous that I would certainly leave out more than one if I listed only a few. Working in science is a privilege, because those of us who have been so fortunate work among some of the most honest, humanistic people to be found anywhere. I must also mention a special friend from the early days, a scientifically oriented, retired senior chemical engineer with the Monsanto organization back in St. Louis. Dan Steinmeyer has always helped me keep my feet on the ground, especially when the excesses of "mature capitalism"—which I often excoriate in this book—become the subject of a discussion. The view from nowhere means looking everywhere and avoiding all ideologies. Good friends who have had different life experiences are needed for that.

The final person I must thank profusely is the one who has shared over fifty years of marriage with me and has been a constant partner through the many challenges and adventures of a long, enjoyable life. Elizabeth is a fine writer in her own right and the best practical psychologist I have ever met. I have benefited for decades having a patient and loving copy editor for everything I have written on humanism over the past two decades. If I have also learned to appreciate what smart women could offer to our troubled world today, an idea developed in this book, Elizabeth, you are the one who taught me.

INDEX

Westfield Memorial Library
Westfield, New Jersey

4/13

001 Jor
Jordan, Stuart D.
The Enlightenment vision